Eagan Press Handbook Series

Wheat Flour

William A. Atwell

Cover: Hard wheat flour particle courtesy of Carl Hoseney, R&R Research Services, Inc.; Rapid Visco Analyser courtesy of Foss North America, Inc.; Quadrumat mill courtesy of Bühler AG; cake slice, ©1997 Artville, LLC.

Library of Congress Catalog Card Number: 2001089692
International Standard Book Number: 1-891127-25-X

Printed in the United States of America on acid-free paper

American Association of Cereal Chemists
3340 Pilot Knob Road
St. Paul, Minnesota 55121-2097, USA

About the Eagan Press Handbook Series

The Eagan Press Handbook series was developed for food industry practitioners. It offers a practical approach to understanding the basics of food ingredients, applications, and processes—whether the reader is a research chemist wanting practical information compiled in a single source or a purchasing agent trying to understand product specifications. The handbook series is designed to reach a broad readership; the books are not limited to a single product category but rather serve professionals in all segments of the food processing industry and their allied suppliers.

In developing this series, Eagan Press recognized the need to fill the gap between the highly fragmented, theoretical, and often not readily available information in the scientific literature and the product-specific information available from suppliers. It enlisted experts in specific areas to contribute their expertise to the development and fruition of this series.

The content of the books has been prepared in a rigorous manner, including substantial peer review and editing, and is presented in a user friendly format with definitions of terms, examples, illustrations, and trouble-shooting tips. The result is a set of practical guides containing information useful to those involved in product development, production, testing, ingredient purchasing, engineering, and marketing aspects of the food industry.

Acknowledgment of Sponsors for *Wheat Flour*

Eagan Press would like to thank the following companies for their financial support of this handbook:

American Ingredients Company
Kansas City, MO, USA
816/561-9050

Buhler AG
Uzwil, Switzerland
+41 71 955 11 11

Cargill, Inc.
Wayzata, MN, USA
952/742-4274

CHOPIN S. A.
Villeneuve La Garenne, France
+33 1 41 47 50 88

ConAgra Grain Processing Company
Omaha, NE, USA
877/717-1694

Earthgrains Company
Clayton, MO, USA
314/259-7000

General Mills
Golden Valley, MN, USA
763/764-7768

North American Millers' Association
Washington, D.C., USA
202/484-2200

Eagan Press has designed this handbook series as practical guides serving the interests of the food industry as a whole rather than the individual interests of any single company. Nonetheless, corporate sponsorship has allowed these books to be more affordable for a wide audience.

Acknowledgments

Eagan Press thanks the following individuals for their contributions to the preparation of this book:

Joel Dick, Roman Meal Milling Co., Fargo, ND
Doug Edmonson, The Earthgrains Company, St. Louis, MO
Keith Ehmke, Cargill, Inc., Wayzata, MN
Dan Lewandowski, General Mills, Inc., Minneapolis, MN
Pat McCluskey, Kansas State University, Manhattan, KS
Ron Moline, Bay State Milling Co., Winona, MN
Wayne Moore, Miller Milling, Minneapolis, MN
Mark Stearns, Interstate Brands Corp., Kansas City, MO
Chuck Walker, Kansas State Univesity, Manhattan, KS
Glen Weaver, ConAgra, Inc., Omaha, NE
Marvin Willyard, Kansas State Univesity, Manhattan, KS

Contents

1. Wheat • 1

Historical Perspective
Wheat Types
Growth Regions
General Uses
Kernel Structure
Germination and Growth
Wheat Production Problems
World Wheat Production and Marketing
Logistics
Wheat Standards
Wheat Storage
Factors Important to the Wheat Producer

2. Milling • 15

Historical Perspective
Wheat Sourcing
Dry Milling: process • products
Milling of Soft Wheat, Hard Wheat, and Durum
Flour Storage
Wet Milling
Factors Important to the Miller

3. Composition of Commercial Flour • 27

Fundamental Composition: protein • starch • nonstarchy polysaccharides • lipids
Functionality of the Flour Components: gluten • starch • other components
Flour Enzymes: amylases • proteases • lipases and lipoxygenases • pentosanases • phytase • polyphenol oxidase
Flour Additives: enrichment • oxidants • reducing agents • chlorination • bleaching agents • malt
Nutritional Aspects

4. Wheat and Flour Testing • 47

Tests Exclusive to Wheat: test weight • thousand-kernel weight • kernel hardness • flour yield

Color Tests: Pekar color (slick) test • Agtron color test

Odor Test

Basic Analyses: moisture • ash • protein • pH • enrichment detection • semolina granulation

Flour Performance Tests: farinograph • mixograph • extensigraph • alveograph • gluten washing tests • alkaline water retention capacity • solvent retention capacity profile

Enzyme Analyses: falling number • amylograph/Rapid Visco Analyser

Viscosity Methods

Microbial Assays

Baking Tests

Starch Tests: starch content • starch damage

Near-Infared Reflectance Methods

5. Specifying "Quality" Flour • 67

Quality and Consistency

Communication

Flour Specifications: general information in a comprehensive flour specification • testing procedures • hard wheat products • soft wheat products • durum products

Meeting and Enforcing Specifications

Crop-Year Changeover: water relationships and new-crop flour • gathering information • assimilating the information • disseminating the information • upgrading specifications

6. Products from Hard Wheat Flour: Problems, Causes, and Resolutions • 79

Ingredients: breads • related products

Processing: scaling • dough processing • proofing • baking • cooling

Product Issues: appearance • texture • flavor • shelf life issues

Processing Issues

Fundamental Mechanism of Breadmaking

Troubleshooting

7. Products from Soft Wheat Flour: Problems, Causes, and Resolutions • 97

Ingredients: flour • chemical leavening • shortening • sucrose • egg whites

Formulation: cakes and related products • doughnuts • crackers • cookies • biscuits • pie crusts

Processing: cakes and related products • doughnuts • crackers • cookies • biscuits • pie crusts

Product and Processing Problems: cakes • doughnuts • crackers • cookies • biscuits • pie crusts

Troubleshooting

8. Durum-Based Products: Problems, Causes, and Resolutions • 115

Ingredients and Formulation: pasta • noodles

Processing: pasta • noodles

Product and Processing Issues: pasta • noodles

Troubleshooting

Glossary • 125

Index • 131

Wheat Flour

CHAPTER 1

Wheat

A wheat seed warms, and moisture begins to penetrate its outer layers. The metabolic processes fostering growth are initiated; a shoot emerges, as does a root, and the seed becomes a fledgling wheat plant. If conditions are favorable, the plant continues to grow, resembling its close cousins, the common grasses found in lawns. The plant continues to grow and, eventually, inside the stem, a head of wheat begins to develop. The stem swells, and then the head emerges. It "flowers," initiating the development of new seeds within the head, which ripen from a soft "milky" consistency to kernels that eventually resemble the seed that initiated this entire growth cycle. As the wheat ripens, the plant turns from green to amber and loses moisture.

It is now time to harvest the new wheat kernels by mechanically separating them from the rest of the plant. Perhaps the wheat will be transported to elevators, stored, and transported to larger elevators before the wheat is tempered and milled into flour—or perhaps the kernels will be milled locally. Regardless of where it is milled, this "new-crop" flour will eventually be used to produce a seemingly endless variety of wheat-based products.

As the major ingredient in most such products, flour exerts a major effect on their quality. To be able to effectively use flour to make high-quality products without encountering processing or end-product quality problems requires a thorough working knowledge of all aspects of wheat and flour. It is the objective of this handbook to be a broad-based resource to aid the reader in acquiring this knowledge.

In This Chapter:

Historical Perspective
Wheat Types
Growth Regions
General Uses
Kernel Structure
Germination and Growth
Wheat Production Problems
World Wheat Production and Marketing
Logistics
Wheat Standards
Wheat Storage
Factors Important to the Wheat Producer

Historical Perspective

Clearly, wheat was one of the earliest and most widely grown agricultural crops cultivated by humans. It is generally accepted that wheat originated in the Tigris and Euphrates River Valley and that the cultivation of wheat as a food crop probably began between 10000 and 8000 B.C. Egyptian tombs 5,000 years old have been unearthed that contain hieroglyphics depicting the harvesting and processing of wheat. Chinese records of wheat date back to 2700 B.C. The hardiness of wheat and the variety of food forms that it can take has made it a truly universal part of the human diet. Today, more wheat is produced worldwide than any other grain.

Wheat Types

The first type of wheat cultivated was einkorn, a *diploid* wheat containing seven *chromosome* pairs. Later a 28-chromosome *tetraploid* wheat known as emmer evolved and was cultivated extensively in the Middle East. Durum wheat, which is used today to make pasta, is also tetraploid. Common wheat, which is used to make the wide variety of dough and batter-based products today, is *hexaploid,* having six sets of each of the seven basic chromosomes.

With respect to biological classification, three *species* of wheat are commonly grown today. The first, *Triticum aestivum,* forms the *classes* hard red winter, hard red spring, soft red winter, hard white, and soft white. *T. compactum* includes the club wheats. The third species is *T. durum,* which includes the durum and red durum wheat classes.

Three sets of terms are used to describe most modern wheat types. The first term (i.e., "hard" or "soft") relates to the hardness of the kernel. Hard wheat requires more energy to mill than soft wheat, because each individual kernel requires more force to crush it. The second term (i.e., "red" or "white") relates to the presence or absence of a reddish pigment in the outer layers of the wheat kernel. A visual examination is all that is required to differentiate these two types of wheat. The third term (i.e., "winter" or "spring") generally describes the growth "habit" of the wheat. Winter wheat is planted in the autumn, sprouts in the spring, and is harvested in the summer. It requires a period of below-freezing temperatures before it can form the heads that ultimately contain the wheat kernels. This process is known as *vernalization* (from the Latin for "spring"). Spring wheat does not require cold weather in order to form heads and is generally planted in the spring and harvested in late summer or autumn. All combinations of growing season, color, and hardness are possible. Consequently, three letters (e.g., HRS for hard red spring, SWW for soft white winter, etc.) are used to describe most common wheats.

The wheats making up the two other species commonly grown are distinctly different from the common wheat types described above. Durum wheat does not require vernalization and produces kernels much harder than common hard wheats. Additionally, desirable yellow pigments are not concentrated in the outer layers of the durum kernel but are distributed throughout the entire endosperm. Club wheats are unique in that they are always soft and usually have a low protein content. However, both winter and spring varieties of club wheats exist, as well as red and white varieties.

HRW, HRS, SRW, durum, hard white (HW), and soft white (SW) are the major classes of wheat. Club wheat is grown in such small quantities that it is often included in the soft white class. Within each class are a large number of wheat *varieties.* Each variety is genetically different and is differentiated from others in some observable or measurable characteristic such as yield potential, disease resistance, drought resistance, or some physical attribute of the plant.

Diploid—Describing an organism with two sets of chromosomes.

Chromosome—A body composed of DNA and carrying part of the genetic code for the organism, i.e., the wheat plant.

Tetraploid—Describing an organism with four sets of chromosomes.

Hexaploid—Describing an organism with six sets of chromosomes.

Species—A biological classification below *genus* and above *variety.*

Class—A type of wheat usually designated by hardness, color, and growing season.

Vernalization—A process required for a winter wheat to create wheat heads. The temperature must drop below freezing for this to occur.

Variety—A biological classification below *species.*

Many wheat varieties were initially brought from one area to another with a similar climate. This was the case with the variety Turkey Red, which was transported from the Ukraine to Kansas. However, new varieties are always under development. Traditionally, this has been accomplished with classical breeding programs in which promising varieties are genetically crossed and their progeny evaluated for positive and negative characteristics.

A good breeding program must take all aspects of wheat quality (e.g., yield, climatic tolerance, protein quantity, protein quality, and baking performance) into account to develop a truly superior variety. The process of creating a new variety takes several years. Individual plants of known varieties with desirable traits are first selected. They are then crossed, and the progeny are grown through several generations under controlled conditions to produce enough seed for evaluation. Evaluations are first performed on a small scale in a laboratory to identify any improvements. Varieties exhibiting desirable characteristics in the laboratory are then grown in test plots under actual growing conditions. Wheat varieties from the test plots are evaluated for growth characteristics (e.g., yield and disease resistance), milling quality, and baking quality. Those exhibiting superior characteristics in these field tests are then released to wheat producers for commercial use. Adoption of new varieties by wheat producers is a subjective process and can also take several years. Since the process of developing, testing, and releasing new wheat varieties is so time-consuming, the percentages of each variety grown in a given area change slowly from year to year. An improved variety may take five years or more between initial development and acceptance as a commercial variety for production.

Recently, biotechnological methods have been applied to improve wheat and create new varieties. These techniques often shorten the time needed for development of a new variety. For example, if a gene is known to have an improving effect from earlier testing, it can be incorporated into the target wheat plant and monitored through the growth generations to ensure that it is retained. This does not eliminate the laboratory testing described above for the traditional procedures, but it does eliminate some traditional selection processes required to ensure that desirable traits are maintained. Of course, varieties developed with these techniques are subjected to field tests and must also meet the other requirements described above.

Waxy and partial waxy wheats are examples of wheats that have recently been developed. In normal wheat, about 75% of the starch is in the form of the branched polymer *amylopectin* and 25% is in the form of the linear polymer *amylose*. Three genes regulate the amount of amylopectin in wheat. By controlling these genes, wheat breeders have been able to develop waxy wheat, in which essentially all of the starch is in the form of amylopectin. In partial waxy wheat, not all of the genes are controlled, and an amylopectin amount intermediate between normal and waxy wheat occurs. Waxy and partial waxy wheats are now being evaluated with respect to end-use qualities.

Amylopectin—A branched polysaccharide composed entirely of glucose units.

Amylose—A linear polysaccharide composed entirely of glucose units.

Growth Regions

Wheat is a hardy crop and can be grown under a wide range of environmental conditions. Renowned wheat-growing regions in the world include the Ukraine, Buenos Aires province in Argentina, the lowlands of Europe, the southeastern and southwestern states of Australia, and the Great Plains of the United States and Canada. Because it is grown in so many places in the world, wheat is being sown and harvested at some location at any given time during the year.

However, not all wheats are grown in all environments. The United States may be unique in that every class of wheat is grown within its boundaries (Fig. 1-1). HRW wheat is grown in a wide belt extending from Texas to South Dakota and Montana. HRS wheat and durum wheat are grown in the Dakotas, Minnesota, and Montana. SRW wheat cultivation extends through the Ohio and southern Mississippi River valleys as well as the southeastern states from Virginia to Alabama. Soft white and club wheats are grown primarily in the Pacific northwestern states of Washington, Oregon, and Idaho. Hard white wheat is now gaining more prominence in Kansas and other HRW-producing areas. Within any given region, many varieties of wheat

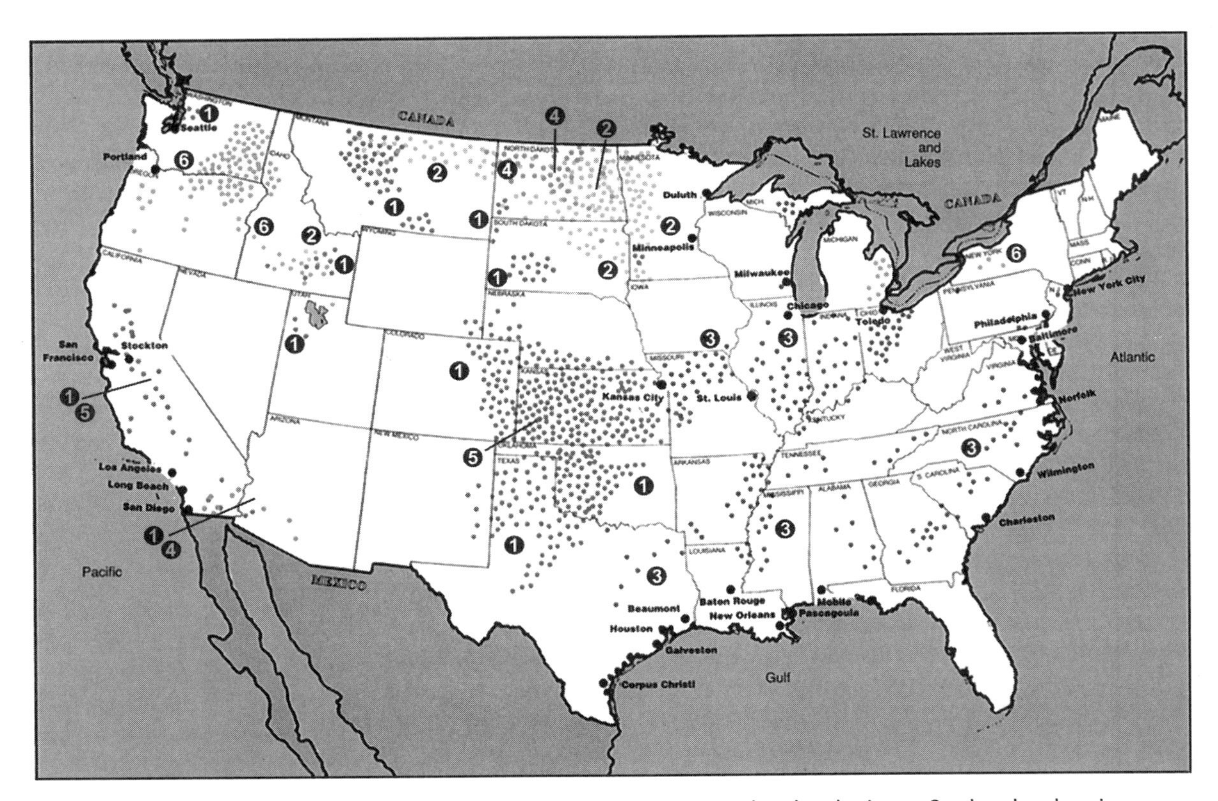

Fig. 1-1. U.S. growth regions for the different classes of wheat. 1 = hard red winter, 2 = hard red spring, 3 = soft red winter, 4 = durum, 5 = hard white, 6 = soft white. (Adapted, with permission, from [1])

may be grown, but a relatively small number of varieties make up the bulk of the acres planted.

General Uses

Not all wheat types are suitable for all products. HRW, HW, and HRS wheat are generally used in the production of breads and related yeast-leavened, dough-based products. This is due in large part to the ability of doughs made from these types of wheat flour to retain leavening gases and subsequently yield breadlike structures and textures. SRW and SW are commonly used in the production of cakes and other batter-based products as well as in crackers and cookies. Flour from this type of wheat generally does not produce highly elastic intermediate or final products. As a result, the textures of products made with soft wheat are generally not very chewy. Soft white and club wheats are used in products such as noodles, where the presence of pigmented bran specks in the product causes an objectionable appearance. Durum wheat is used to produce pasta products such as spaghetti and macaroni. The yellow color inherent in semolina, the primary milling fraction of durum (see Chapter 2), is considered highly desirable and is the standard for pasta products. Table 1-1 summarizes the characteristics and uses of the various wheat classes.

TABLE 1-1. Wheat Classes and Their General Characteristics and Principal Uses

Class	General Characteristics	General Uses
Hard red winter (HRW)	High protein, strong gluten, high water absorption	Bread and related products
Soft red winter (SRW)	Low protein, weak gluten, low water absorption	Cakes, cookies, pastries, pie crusts, crackers, biscuits
Hard red spring (HRS)	Very high protein, strong gluten, high water absorption	Bread, bagels, pretzels, and related products
Hard white	High protein, strong gluten, high water absorption, bran lacks pigments	Bread and related products
Soft white	Low protein, weak gluten, low water absorption, bran lacks pigments	Noodles, crackers, wafers, and other products in which specks are undesirable
Durum	High protein, strong gluten, high water absorption	Pasta

Kernel Structure

Like any seed, the wheat *kernel* is a complex structure with many individual components (Fig. 1-2). However, with respect to processing (i.e., milling), the wheat kernel is divided into three general

Kernel—An individual seed of a cereal grain.

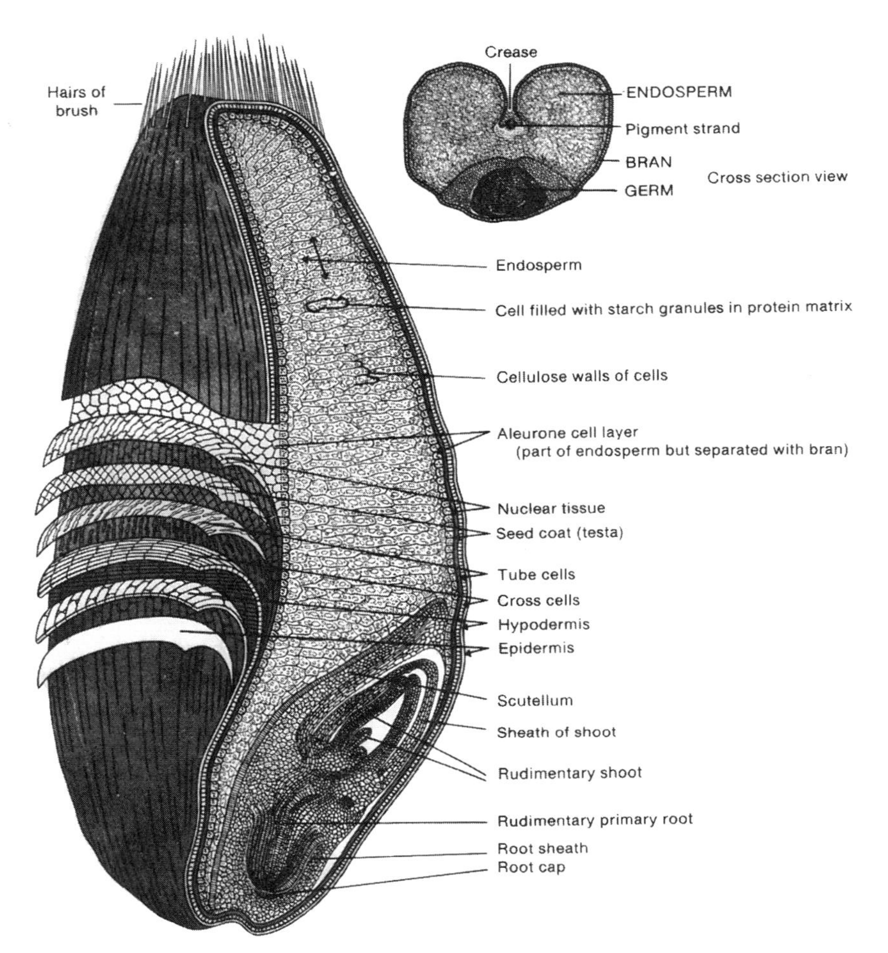

Fig. 1-2. Schematic diagram of the wheat kernel, illustrating the major anatomical parts. (Courtesy Millers National Federation)

anatomical regions. The outer protective layers of the kernel are collectively called the *bran*. The bran constitutes about 14% of the kernel by weight and is high in *fiber* and *ash* (mineral) content. The *germ*, the embryonic wheat plant, constitutes only about 3% of the kernel. Most of the lipid and many of the essential nutrients in the kernel are concentrated in the germ. The remaining inner portion of the kernel is the starchy or storage *endosperm*. It provides the energy and protein for the developing wheat plant. It is characterized by its high *starch* and moderately high protein (i.e., *gluten*) content. The endosperm is the major component of all kernels and is the primary constituent of flour. Finally, a single, highly specialized layer of endosperm cells forms a border between the starchy endosperm and the bran. This layer, called the *aleurone*, is usually considered to be part of the endosperm, but it is biologically much more active and subsequently contains high enzyme activity. Because of its composition, activity, and location, it can exert a variety of negative effects on the acceptability of flour. Consequently, it is generally removed as part of the bran during most flour milling operations, and, in fact, millers consider the aleurone to be part of the bran.

Bran—The outer protective layers of a wheat kernel.

Fiber—Carbohydrates that cannot be digested in the human gut.

Ash—Material, composed primarily of minerals, surviving very high temperature treatment of flour or wheat.

Germ—The potential wheat plant within the wheat kernel.

Endosperm—The major portion of the wheat kernel by weight.

Germination and Growth

Germination is the process that initiates growth of a seed. Like seeds of the other cereals, wheat seeds are dormant and must first be subjected to the appropriate environmental conditions (temperature and moisture) in order to activate hormones within the germ and initiate growth. These hormones, in turn, regulate the production and release of the enzymes governing the metabolic processes involved in growth.

While germination is the required first step in producing a new wheat crop, germination at inappropriate times results in problems for end users of flour. Wet conditions during a harvest can cause the mature seeds to germinate in the field. During germination, enzyme activity, most notably that of *α-amylase*, increases rapidly. If wheat is harvested after it has germinated, it is called *sprouted wheat*, and the flour made from it often creates significant problems in product quality.

Wheat growth is generally divided into four broad stages: *tillering*, *stem extension*, *heading*, and *ripening* (Fig. 1-3). Tillering involves the production of individual wheat plants. Each plant sends out shoots and creates new plants (i.e., tillers). The stem extension phase is further divided into the *jointing* and *boot* substages. Joints (i.e., nodes) in the stem become clearly visible as the stem elongates during jointing. Most wheat grown today is semidwarf wheat, which grows to a height of about 2 ft (61 cm) at maturity. Semidwarf wheat was bred to minimize the number of stems and leaves requiring removal during harvesting. After the second and final joint appears, the developing wheat head swells in the stem, creating what is called the "boot."

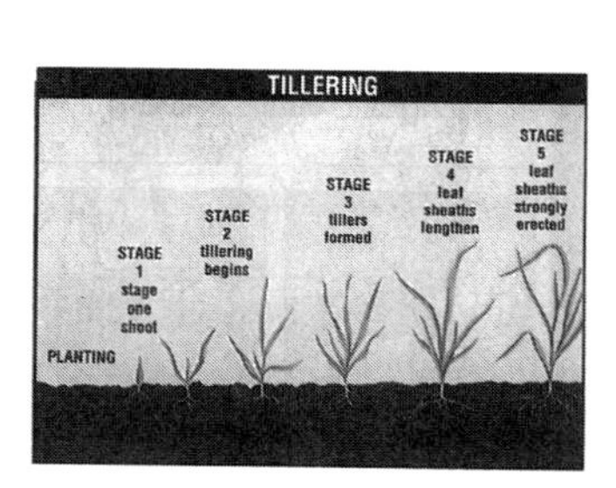

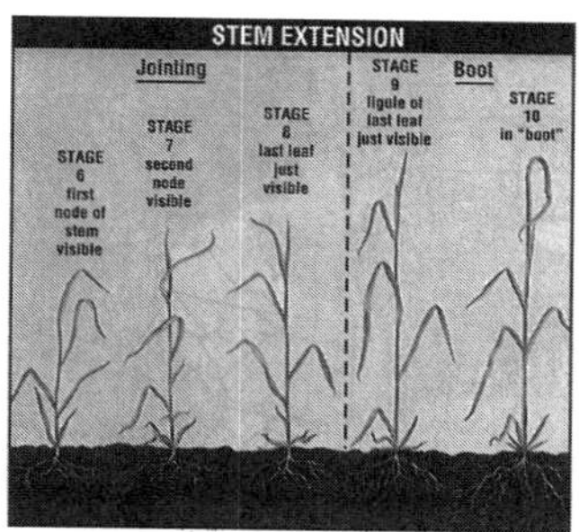

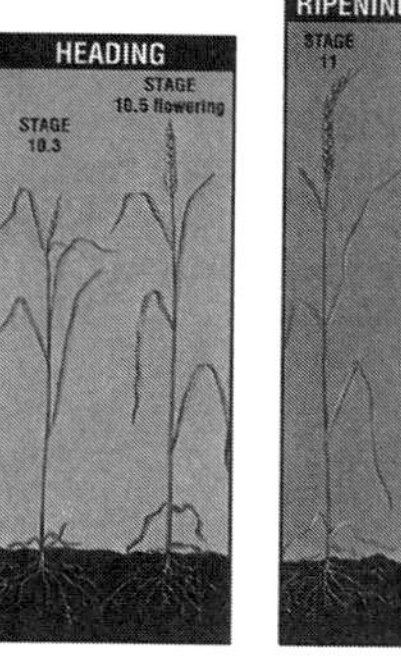

Fig. 1-3. Feekes Scale of wheat development. (Reprinted, with permission, from [1])

The wheat head emerges from the boot to initiate the heading phase. Contained within each wheat head are multiple *stamens* and *pistils*. Flowering occurs during this stage; the stamens pollinate the pistils, thus initiating the development of numerous individual wheat seeds (i.e., kernels) within each head. Finally, during the ripening stage, the seeds fill, becoming harder and drier as the process continues. The seeds are mature but not storage stable before they are dry. At the end of the ripening stage, the wheat is ready to harvest.

The time from germination to harvest can vary with the type of wheat and the growing conditions, but it is generally about three months for spring wheat. It is longer for winter wheat because of its dormancy during the winter months. For HRW and HRS wheat, the average time between the emergence of the heads and harvest is about one month.

Starch—The primary carbohydrate in the wheat endosperm, composed entirely of glucose units linked together.

Gluten—The primary protein complex found in wheat.

Aleurone—The outermost layer of endosperm cells, biologically active and with a high enzyme content.

Germination—The process whereby the wheat seed begins to grow.

α–Amylase—An enzyme that severs the α-1,4 bonds between glucose units in starch.

Sprouted wheat—Wheat that has germinated.

Tillering—The first stage in the development of the wheat plant, initiated when the shoot breaks the ground.

Stem extension—The phase of wheat growth following tillering, in which the stem elongates and the wheat head begins to develop within the stem.

Heading—A stage in the growth cycle of a wheat plant initiated when the wheat head emerges from the boot.

Ripening—The final stage in the growth cycle of wheat, in which seeds mature.

Jointing—The first portion of the stem-extension phase of wheat growth.

Boot—A swelling in the wheat stem caused by the developing wheat head.

Stamen—The male sex organ of plants.

Pistil—The female sex organ of plants.

Yield—The amount of wheat produced per area of land.

Wheat Production Problems

Wheat production involves planting, fertilizing, controlling weeds and pests, irrigation (in some cases), and harvesting. Problems arising during the production of wheat can affect characteristics of the flour that are important to the end user. Weather, insect pests, molds, and weeds can all affect the quality and quantity of wheat produced, and the flour produced from such wheat can vary dramatically as a result.

A prime example is a well-established relationship that exists between *yield* (bushels per acre) and the protein content of the wheat kernel. As more wheat is produced on the same area of land, the protein content of the kernel drops because of the limited available nitrogen in the soil and the number of plants requiring it for growth. Therefore, conditions that affect yield also affect protein content. When temperatures drop below freezing at critical growth stages (e.g., flowering), yields can be compromised; similarly, low-moisture growing conditions often diminish yield. Either, or both, of these conditions can radically decrease yield and lead to wheat with high protein content. Because protein content can have a significant effect on processing or final-product quality, the resultant flour may process poorly or may not produce an optimal product. Conversely, too much moisture during the growing season gives very high yields and depressed protein levels, leading to other, but equally serious, processing or product quality problems. The specific effects of protein level on processing and product quality are discussed in more detail in Chapters 6 and 7.

Wet conditions can cause other problems, of which sprouting is the best known. When wheat seed germinates, the activity of many enzymes rises in the germ and aleurone. Some of these enzymes, if present in the flour, can cause problems such as excessive browning or sticky textures in bread products. Wet conditions also foster the growth of molds (i.e., fungi). Molds can hinder the development of wheat kernels and, in some cases (e.g., infection by *Fusarium*), produce toxins (e.g., vomitoxin) that render the wheat unacceptable for animal or human consumption.

Various insect pests (e.g., weevils, moths, borers, and mites) attack portions of the wheat plant and reduce its ability to develop sound wheat kernels. These pests can also affect wheat in storage by diminishing its inherent quality or by requiring it to have additional processing to remove contamination. Finally, weeds growing in wheat fields can compete for nutrients and affect yield. The weed seeds are often carried along when the wheat is harvested and must be removed before milling (see Chapter 2).

Perhaps the most pressing problems for end users of flour come from variation within a wheat crop caused by environmental factors. Rainfall, farming practices, and soil conditions usually vary significantly over any region where wheat is grown. Although the quality of wheat crops over an entire region may be similar one year, they can vary significantly by location the next year as a result of specific changes in the local environment. It is important that the end user understand this concept. Flour is a complex biological product and is therefore subject to genetic and environmental influences.

World Wheat Production and Marketing

Wheat is grown in almost every country in the world. The United Nations Food and Agricultural Organization (FAO) estimates that, in 1999, over 583 million metric tons (1.3 trillion pounds) of wheat were produced on 215 million hectares (531 million acres). China produces more wheat than any other nation but is generally among the largest importers of wheat as well. Other major producers of wheat include India, the United States, the Russian Federation, the European Union, Canada, Argentina, and Australia. Table 1-2 shows wheat production by continent.

As reported by the Foreign Agricultural Service of the U.S. Department of Agriculture, in 1999, major wheat exporters included the United States, Australia, Canada, the European Union, and Argentina, while major importers included Iran, Brazil, Egypt, Japan, and the Russian Federation (Table 1-3). The United States as an exporter faces very significant competition in the world market. For example, Japanese processors prefer the white wheat grown in Australia for noodle making. Another issue likely to affect U.S. wheat exports is that many wheat users are reluctant to buy genetically enhanced grains. New genetically enhanced varieties are in development and will be available soon.

Wheat is a commodity and, as such, is traded openly in markets throughout the world. Much of the trading occurs in trading organizations located in major cities in wheat-producing areas. Activity at a typical board of trade resembles that of the New York Stock Exchange, with prices projected on screens and buyers and sellers haggling over

TABLE 1-2. World Wheat Production by Continent[a]

Wheat Production (Mt)	Year				
	1996	1997	1998	1999	2000
World	583,675,752	613,133,303	592,342,030	584,697,385	580,014,595
Africa	22,057,403	14,941,088	18,799,724	14,883,412	13,918,414
Asia	243,529,195	266,386,858	254,334,868	260,432,815	249,788,315
Europe	177,882,145	196,530,182	183,872,358	173,689,074	184,247,856
North and Central America	95,183,598	95,496,594	96,650,080	92,504,579	90,620,100
Oceania	23,201,397	19,541,120	21,424,120	24,378,120	22,570,120
South America	21,822,014	20,237,461	17,260,880	18,809,385	18,869,790

[a] Source: Database of the United Nations Food and Agricultural Organization (FAO)—www.fao.org.

TABLE 1-3. Major Importers and Top 12 Exporters of Wheat (July/June year, thousand metric tons)[a]

	1997/98	1998/99	1999/00
Exports			
Argentina	9,606	8,700	10,000
Australia	15,444	16,000	18,500
Canada	21,325	14,455	18,500
India	41	0	200
Kazakhstan	3,428	2,280	4,500
Syria	796	700	100
Turkey	1,274	3,000	1,500
European Union	14,196	14,589	15,500
Eastern Europe	3,098	3,769	2,300
Others	5,964	7,836	4,710
Subtotal	*75,172*	*71,329*	*75,810*
United States	28,090	29,035	29,000
World total	*103,262*	*100,364*	*104,810*
Imports			
Algeria	5,221	4,200	4,500
Brazil	5,969	7,290	6,700
Egypt	7,156	7,430	6,000
India	2,344	1,092	1,600
Indonesia	3,664	3,075	3,200
Iran	3,587	2,538	7,000
Japan	6,200	5,883	5,900
South Korea	3,917	4,868	3,500
Russia	3,085	2,500	4,800
European Union	3,858	3,761	3,600
Eastern Europe	1,932	2,173	2,250
United States	2,488	2,850	2,400
Subtotal	*49,421*	*47,660*	*51,450*
Others	51,663	51,007	51,816
Unaccounted for	2,178	1,697	1,544
World total	*103,262*	*100,364*	*104,810*

[a] Source: Database of the U.S. Department of Agriculture (USDA) Foreign Agricultural Service (FAS)—www.fas.usda.gov

deals in wheat "pits." Often, *hedging* can minimize the risks caused by the volatility of wheat price fluctuations. Hedging involves the use of *futures contracts*, which set the price of wheat to be delivered at some time in the future. In the United States, HRW wheat is traded at the Kansas City Board of Trade, SRW wheat at the Chicago Board of Trade, and HRS wheat at the Minneapolis Grain Exchange. Clearly, price is driven largely by supply and demand. Market fluctuations can be influenced by the size of an incoming wheat crop, conditions under which it is grown or harvested, the amount of wheat carried over from a previous crop, milling requirements, government policies, and economic factors such as the strength of currencies. Consequently, large purchasers and processors of wheat closely monitor worldwide growth conditions, crop quality, and the other factors mentioned above as a means to help ensure that an appropriate supply can be obtained at an appropriate price.

Logistics

In many countries in the world, wheat is harvested and milled to flour in close proximity to its growth location. In other countries, the grain flows through a complicated system of transport, collection (aggregation), and storage before it is finally processed to flour. In these cases, there is usually less *identity preservation*. The wheat arriving at a mill is not of one variety or from a single production area; instead, it is a blend of varieties grown on many farms and in many different environments. Classes of wheat are generally not mixed, but it is often virtually impossible to determine the varieties in a wheat blend.

As an example, in the United States, a producer likely grows only one or two varieties of wheat. Once harvested, the wheat may be stored on site and later brought by truck to a "country" elevator for further storage or sale. At this point, it is mixed with wheat from other farms in the area. This wheat may then be transported to larger

Hedging—A procedure using futures contracts to minimize economic risks caused by fluctuations in wheat market prices.

"terminal" elevators, where it is mixed with wheat from other sources and stored before transport to mill elevators. In most cases, the wheat buyers at a mill will "source" (i.e., procure) wheat by general location and primary attributes such as protein level.

As new varieties with specific end-use characteristics are developed, released, and grown, preserving the varietal identity of wheat (i.e., keeping it segregated by variety) will become more important. Means of segregating wheat varieties will be required, and this will require significant changes in the current supply system. A system involving two ways of handling wheat may evolve, in which "commodity" wheat is handled as it is currently and "specialty" wheat is segregated from other wheat. It will be important that the characteristic being segregated in the specialty wheat be economically advantageous so that it will offset the higher cost associated with a system capable of identity preservation.

Futures contract—An agreement between a seller and buyer in which a price is negotiated in advance for wheat that will be delivered at some point in the future.

Identity preservation—The segregation of wheat varieties to allow them to remain uncontaminated with other varieties between production and end-product use.

Wheat Standards

Because there are so many types of wheat and the potential for mixing them exists, standards for categorizing wheat types have developed. In the United States, all wheat must conform to the wheat definition described in the U.S. Standards for Wheat:

> Grain that, before the removal of dockage, consists of 50% or more common wheat (*Triticum aestivum*), club wheat (*T. compactum* Host.), and durum (*T. durum* Desf.) and not more than 10% of other grains for which standards have been established under the United States Grain Standards Act and that, after the removal of the dockage, consists of 50% or more of whole kernels of one or more of these wheats.

The eight classes of wheat used in inspecting and grading wheat defined in the *Grain Standards of the United States* are HRS, HRW, SRW, durum, hard white wheat, soft white, unclassed wheat, and mixed wheat. There are subclasses further describing the HRS, durum, and white wheat classes, and every class and subclass is divided into five grades. The grades (U.S. No. 1–5) are based on the purity of the wheat (i.e., percent contamination by other wheat or grains), percent of damaged or defective kernels, and foreign material.

Other major exporters also categorize wheat into classes and grades. In Argentina, there are two major wheat types and four grades. Australia categorizes wheat into seven wheat types from seven areas and three major grades. Canada categorizes wheat into seven classes and 19 grades. Standards vary throughout the world, but all are based on criteria such as wheat type, contamination, and quality parameters.

Wheat Storage

Wheat seeds are alive. Consequently, they respire and, if conditions are appropriate, they germinate. Respiration is important because it keeps the germ alive and hence inhibits processes that lead to deterioration of the grain (e.g., oxidative rancidity). However, respiration produces heat and is also associated with weight reduction during storage, so it is advantageous to keep the respiration rate at a minimum. Factors that influence the rate of respiration are temperature, moisture content, and available oxygen. The moisture content of wheat varies with relative humidity, and for practical purposes, wheat stored during commercial distribution is maintained at 14% moisture or below.

A variety of pests (e.g., rodents, insects, and microorganisms) can also reduce the quality of stored wheat; hence, it is important that the wheat be physically isolated to disallow access. Usually grain is stored in silos or elevators, but modern practices also include the use of airtight bunkers, which can be flushed with carbon dioxide or nitrogen to reduce insect infestation and lower the rate of respiration. Large fans are included that can reduce the moisture level of the wheat to stabilize it further.

Factors Important to the Wheat Producer

A farmer is paid on a per-bushel or per-metric-ton basis. That means the more wheat produced, the more lucrative the operation. In some cases, a premium is paid for protein content or some other quality factor, but this generally does not supply much incentive compared with the incentive to produce a high amount of wheat. Consequently, over the years, wheat breeders have developed wheat varieties with increasing yield as the primary objective. Their efforts in developing drought-resistant and disease-resistant varieties with higher yields have been quite successful. In the United States, wheat yields have risen over the past several decades. Unfortunately, the quality of the flour as perceived by the end user (e.g., baker) has not always improved at the same rate.

Many people are involved in the process of developing, growing, transporting, milling, and processing wheat. As described above, their priorities are not always production of the flour that functions best in a product application. Clearly, it is important to understand what is driving each part of this food chain and to communicate frequently with the people along the chain to ensure that there are no misunderstandings.

Web Sites

United Nations Food and Agricultural Organization (FAO)—www.fao.org

FAOSTAT is an on-line and multilingual database currently containing over 1 million time-series records covering international statistics in the following areas: production, trade, food balance sheets, fertilizers and pesticides, land use and irrigation, forest products, fishery products, population, agricultural machinery, and food aid shipments.

U.S. Department of Agriculture (USDA) Foreign Agricultural Service (FAS)—www.fas.usda.gov

Foreign Agricultural Service (FAS) opens, expands, and maintains global market opportunities through international trade, cooperation, and sustainable development activities, which secure the long-term economic vitality and global competitiveness of American agriculture. FAS monitors and assesses global food aid needs and promotes international agricultural trade policies that provide market access for U.S. agricultural commodities. It maintains an international field structure that includes agricultural counselors, attachés, and affiliate foreign national offices, agricultural trade offices, and a number of agricultural advisors covering several countries around the world. FAS also administers a variety of export promotion, technical, and food assistance programs in cooperation with other government agencies, the private sector, and international organizations.

Kansas City Board of Trade—www.kcbt.com

The Kansas City Board of Trade is a place where buyers and sellers gather to trade commodities—commonly known as a commodity exchange. Specifically at the KCBT, futures and options contracts are traded on hard red winter wheat, the Value Line Index of approximately 1,650 stocks, natural gas, and the ISDEX Internet stock index.

Chicago Board of Trade—www.cbot.com

The Chicago Board of Trade (CBOT), established in 1848, is the world's oldest derivatives (futures and futures-options) exchange. More than 3,600 CBOT members trade 48 different futures and options products at the CBOT, resulting in a 1999 annual trading volume of more than 250 million contracts. Early in its history, the CBOT listed for trading only agricultural instruments—such as wheat, corn, and oats. In 1975, the CBOT expanded its offering to include financial contracts.

Minneapolis Grain Exchange—www.mgex.com

After more than 100 years, the Minneapolis Grain Exchange still provides an auction site, as well as many services for buyers and sellers of grains grown in the Upper Midwest and Pacific Northwest. The Grain Exchange boasts the only authorized market for hard red spring wheat, white wheat, and durum futures and options, trading an average of 20 million bushels daily. It is also the largest cash exchange market in the world, trading a daily average of one million bushels of grain, including wheat, barley, oats, durum, rye, sunflower seeds, flax, corn, soybeans, millet, and milo.

Reference

1. Miller's National Federation. 1997. *From Wheat to Flour*. The Federation, Washington, DC.

Supplemental Reading

1. Evers, A. D., and Bechtel, D. B. 1988. Microscopic structure of the wheat grain. Pages 47-95 in: *Wheat Chemistry and Technology*, 3rd ed., Vol. 1. Y. Pomeranz, Ed. American Association of Cereal Chemists, St. Paul, MN.
2. Halverson, J., and Zeleny, L. 1988. Criteria of wheat quality. Pages 15-45 in: *Wheat Chemistry and Technology*, 3rd ed., Vol. 1. Y. Pomeranz, Ed. American Association of Cereal Chemists, St. Paul, MN.
3. Mattern, P. J. 1991. Wheat. Chapter 1 in: *Handbook of Cereal Science and Technology*. K. J. Lorenz and K. Kulp, Eds. Marcel Dekker, Inc., New York.
4. Orth, A., and Shellenberger, J. A. 1988. Origin, production, and utilization of wheat. Pages 1-14 in: *Wheat Chemistry and Technology*, 3rd ed., Vol. 1. Y. Pomeranz, Ed. American Association of Cereal Chemists, St. Paul, MN.
5. USDA. Official United States Standards for Grain, subpart M, United States Standards for Wheat, sections 810.2201–810.2205. 1999. U.S. Department of Agriculture, Grain Inspection, Packers and Stockyards Administration (GIPSA), Washington, DC.

CHAPTER 2

Milling

Historical Perspective

In This Chapter:

Historical Perspective
Wheat Sourcing
Dry Milling
Process
Products
Milling of Soft Wheat, Hard Wheat, and Durum
Flour Storage
Wet Milling
Factors Important to the Miller

Milling is simply the reduction of wheat kernels to smaller particles that can be made into more palatable products. In modern times, it involves, more specifically, the separation of the germ and bran from the endosperm and the reduction of the endosperm to flour. As with fermentation, cooking, and many of the other food processes, the origins of milling are lost in antiquity.

The first miller was the first person to put a wheat kernel in his or her mouth and bite. Clearly, this was not a very efficient means of milling, and rough hand implements such as stones fashioned to rough mortars and pestles (Fig. 2-1) evolved about 8000 B.C. There is evidence of *sieves* to separate flour from unwanted parts of the wheat plant as early as 6000 B.C. The first mechanical mills (i.e., querns) employed two horizontally mounted stone disks. Wheat was fed between them, and the disks were manually rotated. Querns evolved to more sophisticated mills, called rotary mills, in which men and animals powered rotating stone disks cut to more efficiently grind wheat and channel it away. About 100 B.C., waterpower was first employed in milling wheat, and in about 1200 A.D., the first windmills were used. The first manufacturing process to be automated was milling; the implementation of the first milling machinery is credited to Oliver Evans in 1785. The concept of gradual reduction, which persists today, originated in the 19th century in Hungary. The sophisti-

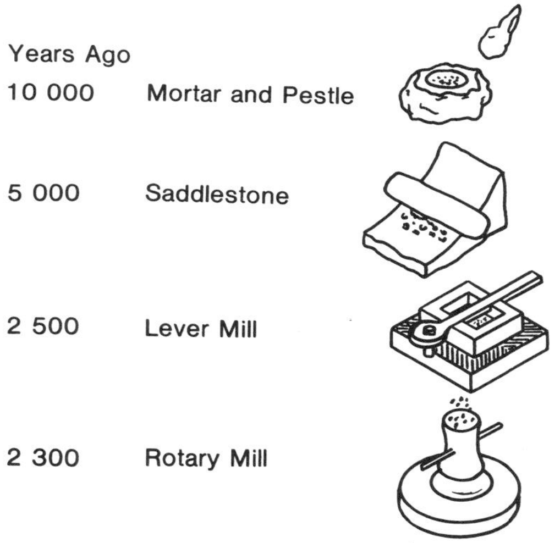

Fig. 2-1. Early grain mills and the time of their introduction. (Reprinted from [1])

Sieves—Devices that separate particles based on size. Smaller particles pass through apertures in the sieve while larger particles are retained.

cated means of grinding, separating, and conveying used in modern mills are based on the same basic processes employed in these early mills.

Wheat Sourcing

A modern miller likely has many customers demanding flours that must meet many different specifications. Even if the mill grinds only one class of wheat (see Chapter 1), which is usually the case, customers require flour characteristics specific to their own applications. That necessitates using different wheat blends and milling practices for each specification. Most mills have the flexibility to store wheat from different sources and to separate and combine different millstreams in a multitude of ways. Some mills also possess the capability to blend flour as an additional means of meeting customers' needs.

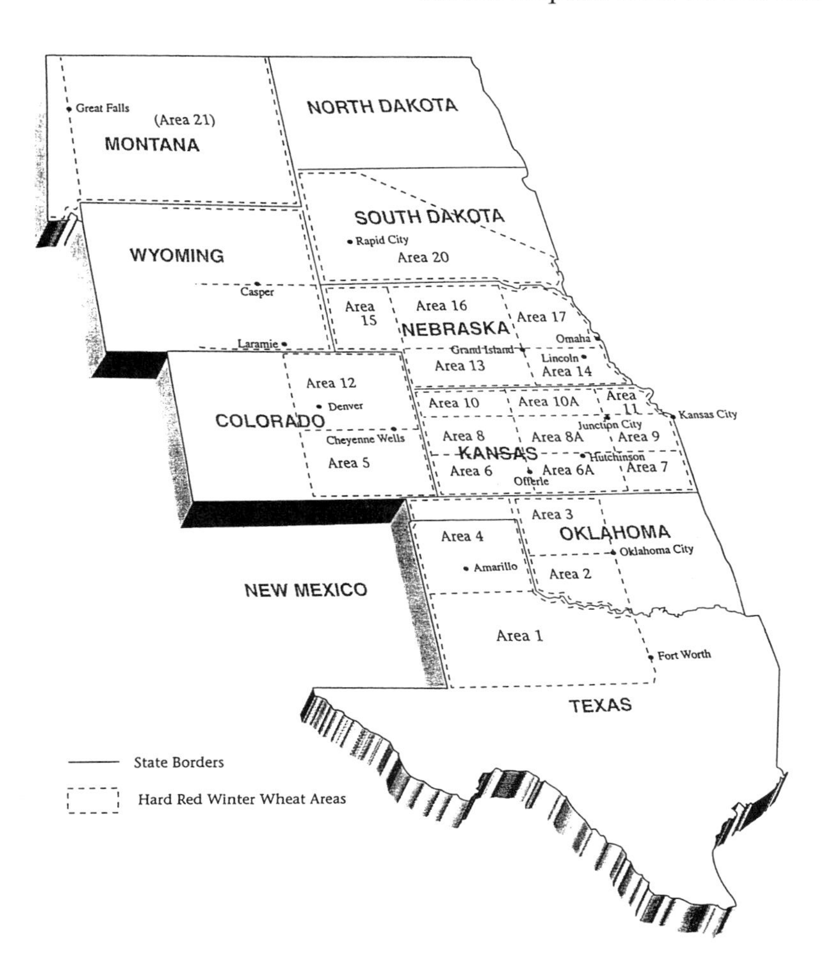

Fig. 2-2. Quality sampling regions for hard red winter wheat. (Courtesy CII Laboratory Services, Inc., Kansas City, MO)

Each year during harvest, much data are collected concerning the incoming crop. In the United States and many other countries, wheat-growing regions are divided into areas for purposes of organizing these data. For example, the HRW wheat region of the midwestern United States is divided into 21 areas (Fig. 2-2), and the HRS wheat region is divided into 10 areas. Wheat harvested in these areas is sampled, subjected to some standard wheat quality tests (e.g., protein, moisture, test weight, 1,000-kernel weight, and flour yield), and milled on small mills in laboratories. The resultant flour is evaluated using a number of standard testing procedures (e.g., protein, moisture, ash, wet gluten, falling number, farinogram characteristics, baking quality, and sedimentation). (These tests are discussed in detail in Chapter 4.) When all of the wheat areas have been harvested and the wheat and flour analyzed, a map of the wheat-growing region containing the results of all the major wheat and flour tests is generated. This information is invaluable to a miller who is *sourcing* wheat. It

Sourcing—The process of identifying and procuring a specific type of wheat or flour.

often gives a competitive advantage and is therefore considered confidential.

Protein level (i.e., amount) is a major factor affecting many customers' requirements. High-protein wheat (e.g., HRS wheat with more than 13% protein or HRW wheat with more than 11%), which generally produces superior bread products, often costs more than wheat with lower protein levels. If a mill's customers require flour with protein ranging from 10 to 12%, the miller will source wheat with the right protein to make both the highest-protein and lowest-protein flours. Customers' specifications requiring intermediate levels of protein will be met by blending the wheat, and possibly the flours, in the proper proportions.

Cleaning—The process of removing unwanted material (i.e., dockage) from wheat before tempering and milling.

Tempering—The process in which water is added to wheat and allowed to equilibrate to toughen the bran and soften the endosperm.

Dry Milling

PROCESS

A modern milling operation involves much more than grinding wheat to a powder. Three general operations are usually involved: *cleaning, tempering,* and *milling*. Cleaning removes unwanted material; tempering softens the grain, making it easier to separate and grind; and milling involves grinding the wheat and isolating wheat components of a specific size. The U.S. government maintains official grain standards that define many of the terms and parameters (e.g., *dockage*, foreign material, damaged kernels, test weight, and moisture) relevant to milling.

Cleaning. Wheat unloaded from a truck, rail car, or ship and conveyed into a mill elevator contains a sizable percentage of nonwheat kernel components, termed "dockage." Dockage consists of other types of seeds, underdeveloped or unsound wheat kernels, insects, stems, stones, and other debris commonly found in a wheat field. Before milling, this debris must be removed, and this is accomplished in the wheat cleaning section of the mill (Fig. 2-3). Although numerous machines exist to clean wheat, they are all classified based on separation by size, shape, density, and magnetism.

Different mills vary greatly with respect to the order of the operations in a cleaning process. Usually, one of the first separations removes any ferrous metal in the wheat using

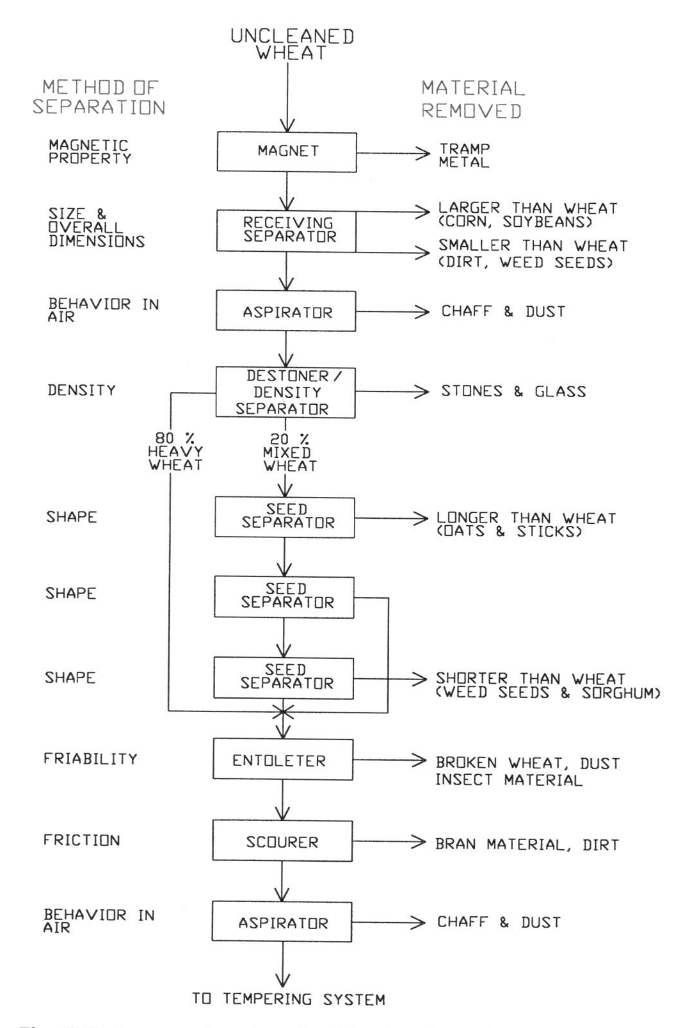

Fig. 2-3. Process steps involved in cleaning wheat before milling. (Courtesy Jeff Gwirtz)

Milling—Grinding of wheat. In a larger sense, all aspects of the conversion of wheat to flour, including cleaning, tempering, grinding, sieving, purifying, etc.

Dockage—Unwanted material in wheat coming into a mill, principally insects, stones, straw, and other contaminants.

Magnetic separator—Device that removes tramp metal during wheat cleaning.

Milling separator—A machine that cleans wheat by drawing lighter, less-dense materials away from wheat in a current of air.

Aspiration—The process of using circulating air to separate materials of different densities.

Disk separator—A wheat-cleaning machine having cavities in rotating disks that exclude or accept grains based on size.

Scourer—A wheat-cleaning machine that removes molds and dirt adhering to wheat kernels using an abrasive screen or surface.

Gravity table—A vibrating inclined plane that cleans wheat by separating materials based on density.

Dry stoner—A machine that utilizes air to separate materials of different densities.

Conditioning—A process in which the equilibration of wheat to a higher moisture level during tempering is facilitated by heat.

magnetic separators. Removing metal early in the process helps avoid damage to equipment farther downstream. A *milling separator* may be next, to remove sticks, stones, stems, and other plant debris. Lighter, less-dense components in the wheat are removed here via *aspiration*. Air circulates upward through the grain as it is fed into the separator, and lighter material is drawn away from the wheat kernels. The wheat then falls onto a sieve, which allows the wheat to pass through but retains stones and larger seeds. Another sieve follows, which retains the wheat and allows smaller seeds to pass through.

A *disk separator*, which separates wheat from other grains of equal density, is also likely to be included in the cleaning process. This machine separates grains based on shape. Pockets in rotating disks accept seeds of certain lengths and reject those of other sizes. Generally there is more than one disk separator. One will accept wheat and another will reject wheat to remove both larger and smaller grains. Dirt or mold adhering to wheat kernels is largely removed using a *scourer*. This machine uses a screen or an abrasive surface to remove any material adhering to the kernel. Materials such as small stones similar in size to a wheat kernel are separated based on density in a *gravity table* or *dry stoner*. The gravity table is an oscillating inclined plane. Denser material such as stones moves down the table faster than lighter material. The dry stoner removes the dense material with aspiration sufficient to raise the grain and allow the stones to drop out.

Tempering. Tempering is the addition of predetermined amounts of water to wheat during specific holding periods. It toughens the bran, making it easier to separate from the endosperm and germ. It also softens the endosperm, allowing it to break apart with less force. Tempering involves adjusting the moisture level of the wheat. For soft wheat, optimal tempering brings the grain to 13.5–15.0% moisture and takes 6–10 hr. For hard wheat, the final moisture is 15.5–16.5%, and tempering times are 12–18 hr. Incoming wheat is generally lower in moisture content than this; hence, water is usually added and the grain is allowed to equilibrate for a period of time. This time varies considerably based on the hardness of the wheat. *Conditioning* of wheat refers to the application of heat in the tempering process to increase the rate of penetration of moisture into the kernels. Temperatures lower than 50°C are employed during conditioning to ensure that the functionality of the flour components, especially the gluten, is maintained.

Milling. At this point, the wheat is ready for milling and starts through the various systems in the mill (Fig. 2-4). The first machine in almost every mill is the *roller mill*. Two rolls, one rotating clockwise and the other counterclockwise, are separated by a small distance called the "gap." One of the rolls usually rotates faster than the other one. Consequently, at the nip, the rotation of the rolls is in the same direction and the wheat experiences a shearing action as well as a crushing action.

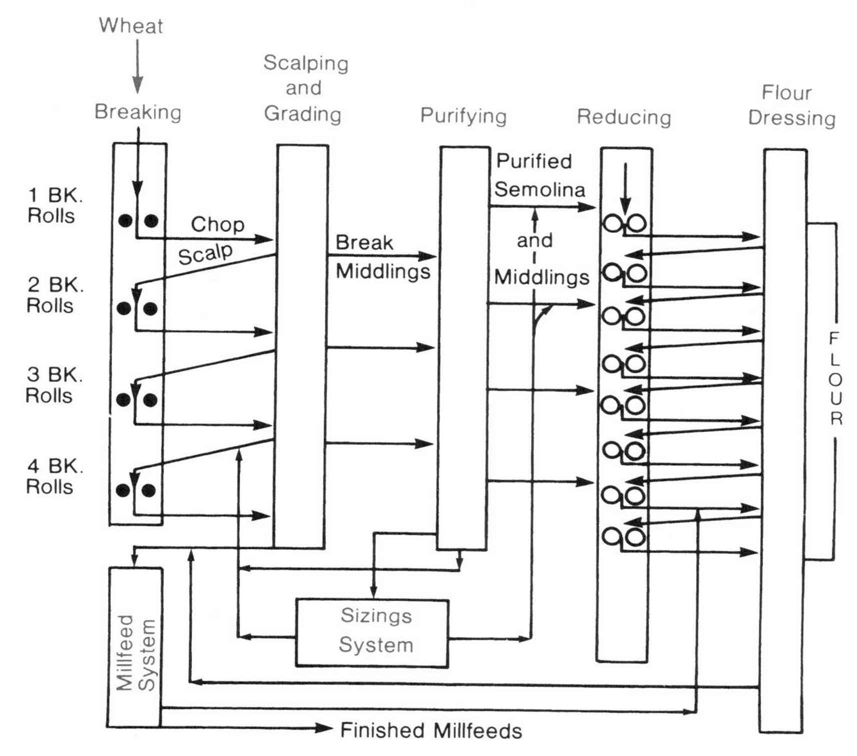

Fig. 2-4. Highly simplified flow diagram for a flour mill, showing the major systems (break, sizing, reduction, and feed) as well as the flow of product through and between the systems. (Reprinted from [1])

Roller mill—The machine in a mill that grinds wheat, as well as wheat particles from roller mills earlier in the process.

Break system—The initial process in wheat grinding.

Purifiers—Machines that remove light, low-density material from higher-density particles using aspiration.

The first roller mills are employed in the *break system*. This is the part of the milling operation designed to remove the endosperm from the bran and germ. Rolls in this process have spiral grooves called "corrugations" cut parallel to the long axis of each roll (Fig. 2-5). Generally there are about five roller mills or five "breaks" in the system. The germ is removed in the first two breaks, as is much of the bran. The germ is pliable and tends to flatten when it goes through the rollers. Bran particles are usually in the form of low-density small flakes. These properties allow millers to separate the germ and bran fractions from the endosperm fraction. After each break, a set of sieves (Fig. 2-6) and/or *purifiers* (aspirators) (Fig. 2-7) separates the ground material by size and density. Small particles are channeled into the

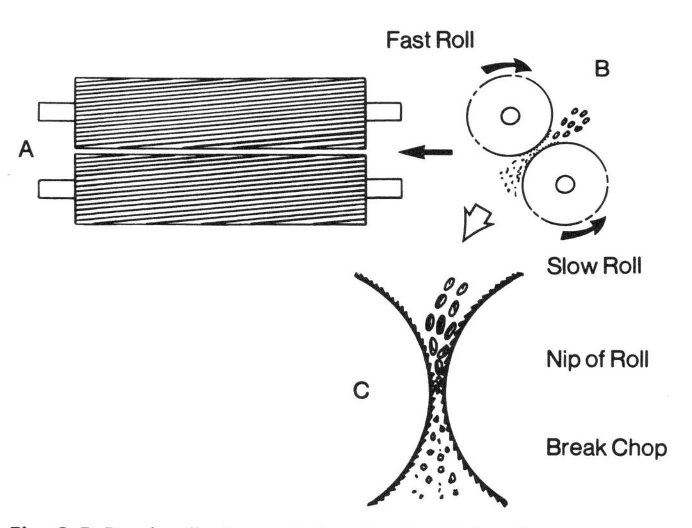

Fig. 2-5. Break rolls shown in longitudinal (A) and cross (B) section. Corregations are visible in A; milling action of the rolls is shown in B and C. (Reprinted from [1])

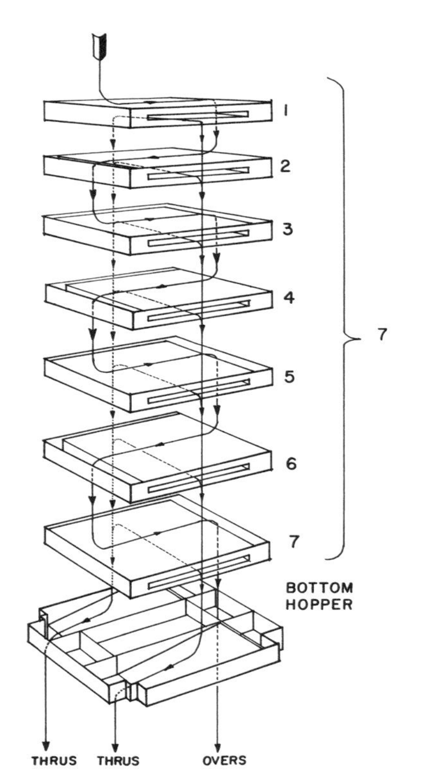

Fig. 2-6. Schematic diagram of the sieves (1–7) in a small sieve box, showing the flow of product. (Courtesy Allis Chalmers)

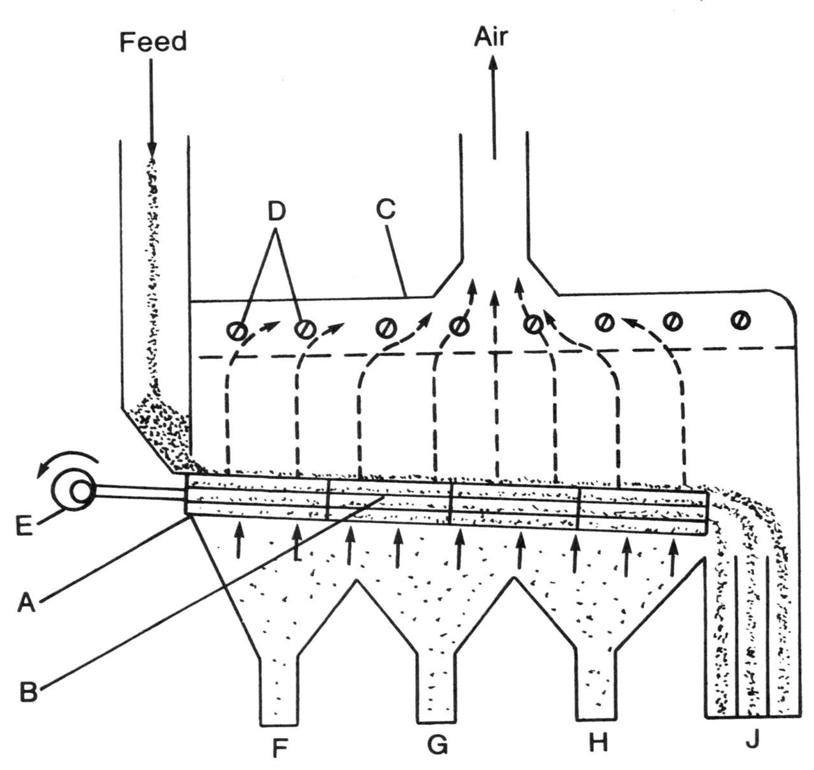

Fig. 2-7. Schematic diagram of a purifier, including the flow of air and product through it. A = frame; B = sieves, C = air trunk, D = flow control valves, E = excentric driver, F = exit for dense (endosperm) particles, G and H = exits for composite (bran plus endosperm) particles, J = exit for bran particles. (Reprinted from [1])

Middlings—Large particles of endosperm obtained after the break system in a mill.

Reduction system—A milling process that employs roller mills with smooth rolls to reduce particle size.

Flour dressing system—The final sieving of flour through 10XX bolting cloth (i.e., a sieving material with apertures of 136 µm).

Chlorine—A gas often applied to soft wheat flour to improve cake-making ability.

flour, and large particles are either removed (as is the case with the germ and bran) or sent to the next break (as occurs for large endosperm pieces).

Once the endosperm is isolated, the large particles that result (called *middlings*) are reduced in the *reduction system* to a particle size distribution consistent with flour. This means they must be able to pass through a 136-µm opening. The rollers in the reduction system are smooth and are operated at low differentials, providing a crushing action that yields the fine particles of a flour (although a small amount of shear is still important). A large percentage of the particles composing the final flour comes off the reduction rolls.

Flour from the break and reduction rolls may be combined in many ways to create different types of flour, but it is usually sifted again in the *flour dressing system*. Material that passes through the 10XX sieves in this process meets the particle size standard for flour. Larger particles are recirculated back to the appropriate point in the grinding process. The flour may be further treated with *chlorine* or supplemented with *enrichment, malt,* and/or a *bleaching agent* depending on the requirements of the customer. Flour additives are discussed thoroughly in the next chapter.

In the *millfeed system,* the germ and bran are separated from each other, and adhering endosperm is removed. The coarse bran from the early breaks is termed "bran" and composes about 11% of the total products from the mill. The finer branny material from the later steps is called *shorts;* it represents about 15% of the total. Germ is generally recovered at the rate of about 0.5–2.0% of the total wheat depending on the type of equipment used. These products are usually sold separately as animal feed, specialty products, or ingredients for human consumption.

Enrichment—Nutrients added to flour for nutritional purposes.

Malt—Barley that has been allowed to germinate. It is ground and added to flour to bolster enzymatic activity and/or to improve flavor.

Bleaching agent—A chemical added to flour to bleach pigments and thus whiten flour.

Millfeed system—The process in a flour mill that separates and purifies flour milling byproducts, specifically bran and germ.

Shorts—A very-high-ash product produced in a mill.

Extraction rate—The amount of flour made as a percentage of total wheat ground. Whole wheat has an extraction rate of 100%. A straight-grade flour, which has an extraction rate of 72%, contains much less bran and germ.

PRODUCTS

If the entire wheat kernel is ground, separated, and recombined, the resultant product is called whole wheat flour. The *extraction rate* for whole wheat flour is essentially 100%, because all of the wheat has been recovered as flour (Fig. 2-8). A flour with most of the bran and germ removed, representing about 72% of the kernel (i.e., an extraction rate of 72%), is termed *straight-grade flour. Patent flours* are those from which many of the flour streams containing high bran content have been removed. These flours contain the lowest amount of bran. Their extraction rates are always less than 72%, generally ranging from 65% extraction for long-patent flour to 45% extraction for short-patent flour. Flour produced solely from the fractions between 45 and 65% extraction is termed *cutoff flour. Clear flour* (or low-grade flour) is composed of the flour streams between 65 and 72% extraction. Clear flour is usually dark because these fractions are quite high in bran. Thus, extraction rate is an estimate of the "purity" of the endosperm, or more accurately, its freedom from non-endosperm components. Therefore, extraction rate is a rough first reflection of certain quality aspects because the non-endosperm components can have adverse effects on processing or product quality.

The less bran and germ in a flour, the lower the mineral content because minerals are concentrated in these fractions (Table 2-1). Hence, the ash test is often used to quantify the purity of a sample. High-extraction flour has higher ash content than lower-extraction flour. Similarly, the endosperm is white, whereas the bran and germ are not, so visual tests can also be used to determine the general composition of a flour. These tests are discussed extensively in Chapter 4.

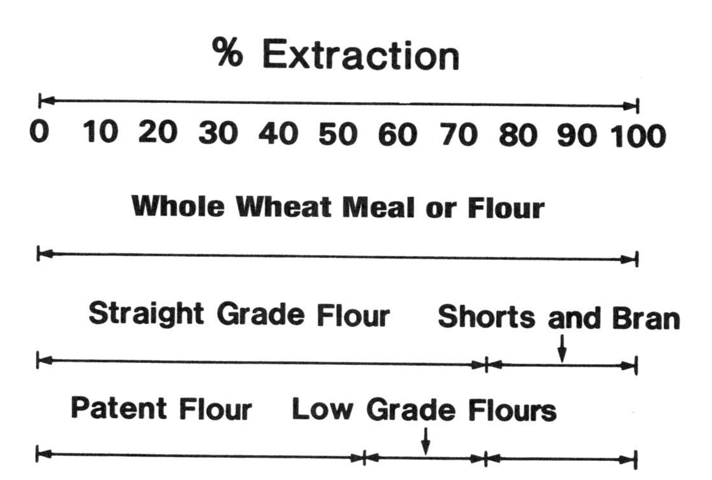

Fig. 2-8. Typical grades of flour produced by flour mills. Whole wheat meal is 100% extraction (second line) but can be subdivided into straight-grade flour and bran plus shorts (third line). Straight-grade flour can be further subdivided into patent and low-grade flours (bottom line). (Reprinted from [2])

TABLE 2-1. Change in Composition from Wheat to Flour[a]

	Wheat	70% Extraction Flour
Ash, %	1.55	0.4
Fiber, %	2.17	Trace
Protein, %	13.9	12.9
Oil, %	2.52	1.17
Starch, %	63.7	70.9
Thiamin, µg/g	3.73	0.70
Riboflavin, µg/g	1.70	0.70
Niacin, µg/g	55.6	8.50
Iron, mg/g	3.08	1.42
Sodium, mg/g	3.2	2.2
Potassium, mg/g	316	83
Calcium, mg/g	27.9	12.9
Magnesium, mg/g	143.0	27.2
Copper, mg/g	0.61	0.18
Zinc, mg/g	3.77	1.17
Total phosphorus, mg/g	350	98
Phytate phosphorus, mg/g	345	30.4
Chlorine, mg/g	39.0	48.4

[a] From (2); used by permission.

Milling of Soft Wheat, Hard Wheat, and Durum

Wheat of the different classes varies significantly in kernel hardness and consequently in how much power is required to grind the grain. Hardness of wheat is a result of the strength of the adhesion between starch and protein in the endosperm. In durum and hard wheat, the interactions between protein and starch are strong, whereas the interactions are weaker in soft wheat. Hence, when the kernel fractures during milling, it breaks apart in different ways. For example, starch occurs in the cells of the endosperm in partially crystalline aggregates called *starch granules*. The space between the granules is filled primarily with the amorphous gluten proteins. When soft wheat is milled, the endosperm cells are converted to flour consisting largely of free starch granules and small particles containing both starch granules and protein. In hard wheat, however, the force holding the starch to the protein may be so strong that, in some cases, the starch granules fracture before the protein-starch interactions are severed. Granules broken in this manner compose what is termed *damaged starch*. In hard wheat, damaged starch may constitute 8% or more of the total starch in the flour. If durum wheat is milled to the particle size of flour, the amount of damaged starch is considerably higher because the interactions between gluten and starch are even stronger.

The description of the milling process given above applies to all types of wheat. Clearly, there are differences in how the process is operated depending on process flow details, the type of wheat, and the particular mill. In general, compared to a hard wheat mill, a soft wheat mill will temper the wheat for a shorter time since the endosperm is already softer. The kernel breaks apart easily so there are usually fewer roller mills in the reduction system. Because the endosperm breaks up more easily, the particle size distribution of soft wheat is smaller and narrower. Thus, there are differences in the sizes of the apertures in the sieves used in soft versus hard wheat mills. The smaller particle sizes of soft wheat often make sifting more difficult because of the tendency of these small particles to attract each other, aggregate, and subsequently not pass through a sieve. This can become a major problem in the operation of a soft wheat mill.

The objective of durum milling is not to reduce the endosperm to flour particle size, but instead to convert it to *semolina*, which has a

Straight-grade flour—The primary product of most flour mills, having an extraction rate of about 72%.

Patent flour—Straight-grade flour with some of the higher-ash components removed. Extraction rates range from 45% (i.e., short-patent flour) to 65% (i.e., long-patent flour).

Cutoff flour—The portion of flour between 45 and 65% extraction.

Clear flour—A high-ash flour fraction consisting of the portion of flour between 65 and 72% extraction. It is also called low-grade flour.

Starch granules—Discrete, partially crystalline aggregates of starch in the wheat endosperm, composed of amylose and amylopectin.

larger particle size. Semolina generally is of a particle size such that no more than 10% will pass through a 180-µm sieve. Specks in the pasta products made from semolina are highly undesirable, and consequently cleaning of durum wheat is quite thorough. Tempering durum wheat generally takes only about 4 hr because it is not necessary to soften the endosperm, which would lead to the production of more flour and less semolina. This short tempering period is sufficient to soften the bran, however, to ensure that it can be removed and separated. The break system is usually more extensive in a durum mill than in a hard or soft wheat mill in order to gradually reduce the grain and avoid the production of flour. The rough semolina from the break system is sent to corrugated sizing rolls in the *purification system* and finally to an abbreviated reduction system to gradually reduce it to a smaller particle size.

Damaged starch—Starch granules that have been physically broken during milling.

Semolina—The primary product of a durum mill, used almost exclusively to make pasta.

Purification system—The part of a durum mill after the break system in which rough semolina is gradually reduced in size.

Flour Storage

Flour is a stable ingredient if it is stored properly. It must be kept dry, at a moisture content of about 14% or below. At higher moisture contents, it is very susceptible to mold growth. If the moisture content becomes too high, microbial growth accelerates and the flour may ferment. High-moisture flour can also cause problems when it is conveyed through a manufacturing facility since it is more prone to clumping and plugging. Flour is porous and as such is prone to absorb chemical compounds. For example, if flour is exposed to the vapors of an organic solvent, it is likely that the solvent's odor will be carried with the flour, possibly even into the final product. High temperatures (i.e., >55°C [>130°F]) for prolonged periods can adversely affect the functionality of the gluten proteins in flour, so holding good flour at elevated temperatures will, with time, cause a reduction in the volume of the bread loaves or cakes baked with it. Another possible mode of deterioration is through insect infestation. Insect eggs can sometimes survive the milling process and infest otherwise good-quality flour. However, low temperatures can hinder the development and activity of insects. Therefore, the flour from a mill that is loaded first into bins and then into bags, trucks, railcars, or ships must be kept dry, at appropriate temperatures, and free of contamination.

Wet Milling

The objective of wet milling is to separate and isolate wheat starch and gluten. The starting material for wet milling is flour, not wheat. The flour is first mixed with water to form a soft or weak dough. In dough, the gluten forms a continuous mass that can be separated from the starch granules by mechanically kneading the dough in a stream of water. Starch, water-soluble materials, and water will pass

Vital wheat gluten—Gluten that is extracted and dried during wet milling.

Weak flour—Flour that produces a less cohesive dough that does not retain gas well.

Strong flour—Flour that produces a very elastic dough that retains gas well.

though a sieve and the gluten will be retained on it. With repeated washing, the gluten can be purified to the point that it is about 75% protein; the starch is further purified in large continuous centrifuges. For the isolation of small quantities of gluten and starch, this process can also be done by hand kneading dough under a stream of water in a sink. In this case, the water-soluble components of the flour are generally not recovered. In the commercial process, the purified starch and gluten products are dried in ring dryers. The drying process must be controlled carefully so as to not affect the functionality of these products. The gluten isolated by this process is called *vital wheat gluten*. It is often used in formulations made with *weak* flour to bolster the volume or attain a more elastic texture in a baked product. Wheat starch is used as an ingredient to dilute the gluten in flours that are too *strong* or in other applications in which more commonly used starches (e.g., corn starch) yield an undesirable flavor or texture.

Factors Important to the Miller

Bakers desire the lowest-cost flour possible to make their product. Similarly, millers strive to reduce costs and maximize profits. Generally, the more flour a miller produces from a given quantity of wheat, the more money the milling process generates. There is incentive, therefore, for a miller to include as much bran in flour as possible while still meeting the specifications of customers. In the United States, there has been a definite trend in the past decades toward higher-ash flours for use in the commercial production of many flour-based products. For example, in the early 1980s, flour used to make bread would often contain 0.40% ash. It is very common today for flour used for this same purpose to contain 0.50% ash. Although

Web Sites

Association of Operative Millers—www.aomillers.org

The Association of Operative Millers (AOM) is an international organization of flour millers, cereal grain and seed processors, and allied trade representatives and companies devoted to the advancement of technology in the flour-milling, cereal grain, and seed-processing industries. The AOM provides an international forum for networking, the exchanging of ideas, technical and educational opportunities, and the discovering of new products and services. Founded in 1896, the AOM was established to improve professionalism and competency in the flour milling, cereal-grain, and seed-processing industries.

World Grain—www.sosland.com/worldgrain

World Grain is the international magazine of grain, flour, and feed.

Grainnet—www.grainnet.com

Grainnet is the online news service of *Grain Journal*, *Milling Journal*, and *Seed Today*.

ash content can vary with variety and growing conditions (e.g., soil composition), this trend is generally not the result of any fundamental change in the mineral content of the wheat grown.

Varieties within a class can vary somewhat with respect to the percentages of endosperm, bran, and germ. These variations affect the amount of flour (of constant ash content) that can be produced from a given amount of wheat. Flour yield, a parameter that is determined on laboratory mills, reflects these variations, and obviously, wheat with a high milling yield is beneficial for the miller. Additionally, a miller prefers wheat with a high test weight and high 1,000-kernel weight. These parameters are indicative of the soundness of wheat. Hardness can also vary, and kernels with hardness of an appropriate degree are desirable to control the amount of starch damage. These tests of milling quality are reviewed in Chapter 4.

References

1. Bass, E. J. 1988. Wheat flour milling. Pages 1-68 in: *Wheat Chemistry and Technology*, 3rd ed., Vol. 2. Y. Pomeranz, Ed. American Association of Cereal Chemists, St. Paul, MN.
2. Hoseney, R. C. 1994. *Principles of Cereal Science and Technology*, 2nd ed. American Association of Cereal Chemists, St. Paul, MN. Chapter 6.

Supplemental Reading

1. Miller's National Federation. 1997. *From Wheat to Flour*. The Federation, Washington, DC.
2. Hoseney, R. C. 1994. *Principles of Cereal Science and Technology*, 2nd ed. American Association of Cereal Chemists, St. Paul, MN. Chapter 7.
3. USDA. Official United States Standards for Grain, subpart M, United States Standards for Wheat, sections 810.2201–810.2205. 1999. U.S. Department of Agriculture, Grain Inspection, Packers and Stockyards Administration (GIPSA), Washington, DC.
4. Posner, E. S., and Hibbs, A. N. 1997. *Wheat Flour Milling*. American Association of Cereal Chemists, St. Paul, MN.
5. Schaetzel, D. E. 1982. Bulk storage of flour. Pages 479-501 in: *Storage of Cereal Grains and Their Products,* 3rd ed. C. M. Christensen, Ed. American Association of Cereal Chemists, St. Paul, MN.

CHAPTER 3

Composition of Commercial Flour

Fundamental Composition

PROTEIN

Protein usually constitutes 7–15% of common flour on a 14% moisture basis (Table 3-1). A common means of classifying the different proteins of wheat and other grains was devised by Thomas Osborne in the early 1900s. The Osborne classification system is based on solubility. Water-soluble proteins, such as many of the enzymes (i.e., biological catalysts) of wheat, are called *albumins*; they make up about 15% of the flour proteins. Many other enzymes are *globulins*, proteins that are soluble in salt solutions. The globulins are relatively minor, making up only about 3% of the total protein.

TABLE 3-1. Analytical Composition of Flour and Its Primary Components[a]

Property	Percent
Moisture	14 (of flour)
Protein	7–15 (of flour)
Osborne classification	
Albumins	15 (of protein)
Globulins	3 (of protein)
Prolamin (gliadin)	33 (of protein)
Glutelin (glutenin)	16 (of protein)
Residue	33 (of protein)
Gluten	6–13 (of flour)
Gliadin	30–45 (of gluten)
Glutenin	55–70 (of gluten)
Starch	63–72 (of flour)
Amylopectin	75 (of starch)
Amylose	25 (of starch)
Nonstarchy polysaccharides	4.5–5.0% (of flour)
Pentosans/hemicellulose	67 (of NSP)
Insoluble	67 (of pentosans/hemicellulose)
Soluble	33 (of pentosans/hemicellulose)
Beta glucans	33 (of NSP)
Lipids	1 (of flour)

[a] Compiled from data in the text.

In This Chapter

Fundamental Composition
- Protein
- Starch
- Nonstarchy Polysaccharides
- Lipids

Functionality of the Flour Components
- Gluten
- Starch
- Other Components

Flour Enzymes
- Amylases
- Proteases
- Lipases and Lipoxygenases
- Pentosanases
- Phytase
- Polyphenol Oxidase

Flour Additives
- Enrichment
- Oxidants
- Reducing Agents
- Chlorination
- Bleaching Agents
- Malt

Nutritional Aspects

Albumins—Proteins that are soluble in water.

Globulins—Proteins that are soluble in salt solutions.

Prolamins—Proteins that are soluble in aqueous alcohol.

Gliadin—One of the two major components of gluten. It is composed of individual protein chains and imparts the viscous nature to gluten.

Glutenin—One of the two major components of gluten. It is a very large molecule that consists of protein chain subunits connected by disulfide linkages.

Glutelins—Proteins that are soluble in dilute acids.

Amino acids—The building blocks of proteins. Each contains an amine group, a carboxylic acid group, and a side group.

Peptide bonds—The type of bond that forms when the amine group of an amino acid reacts with the carboxylic acid group of another amino acid.

Primary structure (of a protein)—Its sequence of amino acids.

Prolamins are cereal proteins generally soluble in 70% aqueous ethanol. *Gliadin*, one of the two major components of the wheat gluten complex, is a prolamin; it constitutes about 33% of all the proteins in flour. The other major component of gluten, *glutenin*, is classified as a glutelin. *Glutelins* are proteins that are generally soluble in dilute acids or bases. Glutenin accounts for about 16% of the flour protein. The Osborne classification of proteins is helpful in that proteins with significantly different properties may be isolated based on their solubility in the solvents above. In practice, however, the separations are not absolute. For example, wheat protein, if treated successively with these solvents, does not totally dissolve. Some wheat protein defies dissolution in any of these solvents; hence, an unclassified residue, which can account for 33% of the total protein, is always left.

Amino acids are the building blocks of proteins. Twenty of them occur naturally in most proteins, and each contains an amino group, a carboxylic acid group, and a side group (referred to as the R group). The amino groups and carboxylic acid groups are bound together in proteins to form *peptide bonds* and consequently string amino acids together to form long protein chains. The sequence of amino acids in the protein chain is called the *primary structure* of the protein. The R group is not involved in the peptide bond. However, the character of the R group influences how the protein interacts with other protein chains or other constituents in the system. Figure 3-1 shows the structure of some amino acids common in wheat proteins. They are linked with peptide bonds.

The amino acid composition of the gluten proteins, gliadin and glutenin, is somewhat remarkable in that relatively few amino acids

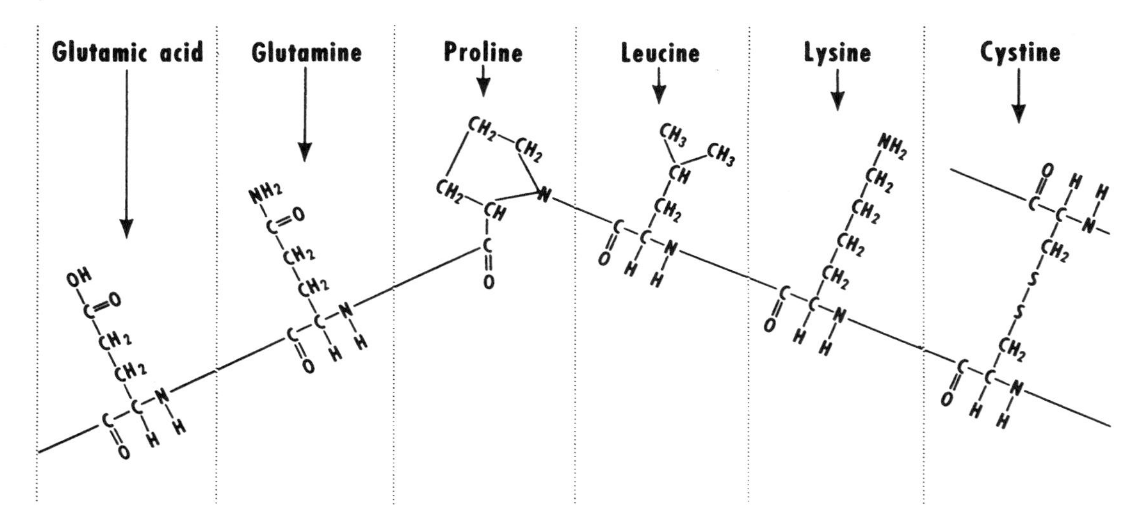

Fig. 3-1. Schematic diagram of the structure of some amino acids common in wheat, showing linkage with peptide bonds. (Reprinted, with permission, from [1])

predominate. *Glutamine*, an amino acid that contains an amide side group that binds water well, constitutes over 40% of all of the amino acids composing these proteins on a molar basis. Another amino acid composing about 15% of gliadin and 10–12% of glutenin is *proline*, which has a cyclic R group structure that puts a bend in a chain of amino acids. Often, protein chains coil and form helices that are considered a *secondary structure* of the protein. Another secondary structure of proteins is the "pleated sheet," which occurs when chains fold back upon themselves. Proline inhibits the formation of these types of secondary structures in gluten. Although it constitutes only 1–3% of gluten proteins, another amino acid of significance is *cysteine*. Cysteine is unique in its ability to form bonds connecting protein chains with its sulfur-containing R group, which constitutes another type of secondary structure of proteins. The connections are called *disulfide bonds*, and the formation or destruction of them has a major effect on the size of glutenin molecules. These three amino acids (glutamine, proline, and cysteine) play a major role in explaining the characteristics of gluten proteins, but gluten does contain other amino acids as well. They can be characterized into four types depending on the structure of their R group: acidic, basic, neutral hydrophilic, and neutral hydrophobic (Table 3-2). The acidic and basic amino acids enter into interactions involving electrostatic attraction or repulsion, and the neutral amino acids exert an influence on how well the protein binds water.

The *tertiary structure* of a protein involves the three-dimensional structure of the protein as a whole. How the R groups are oriented in space in such a structure dictates how the protein interacts with other molecules in its environment. If the tertiary structure is destroyed (e.g., by heat or shear) the protein is said to be *denatured*. Denatured proteins do not have the same characteristics as native or unaffected proteins, even though the primary and secondary structures are the same.

The molecular weight of gliadin ranges from 30,000 to about 125,000. Gliadin exists as single chains. Disulfide linkages exist, but they link cysteine R groups in the same chain. Because of the high level of proline, only about 20% of gliadin chains exist in a helical structure, and there is little evidence of pleated sheets. The tertiary structure is thought to be compact, with many binding interactions occurring between R groups within gliadin molecules.

Glutamine—An amino acid that makes up about 40% of the gluten protein.

Proline—A cyclic amino acid that creates bends in protein chains.

Secondary structures (of a protein)—Structures involving single chains.

Cysteine—An amino acid with a sulfur-containing side group that can bind to other cysteine groups to form disulfide linkages.

Disulfide bonds—"Bridges" between cysteine residues in the same protein chain or in adjacent protein chains.

Tertiary structure (of a protein)—The way the entire molecule is oriented in space.

Denaturation—The process of destroying the tertiary structure of a protein. Denaturing an enzyme eliminates its activity as a catalyst.

TABLE 3-2. The Amino Acids, Grouped by Charge and Hydrophobicity[a]

Acidic	Basic	Neutral (Hydrophilic)	Neutral (Hydrophobic)
Glutamic acid	Lysine	Glutamine	Valine
Aspartic acid	Histidine	Asparagine	Leucine
	Arginine	Serine	Isoleucine
	Tryptophan	Threonine	Alanine
			Phenylalanine
			Tyrosine
			Cysteine
			Cystine
			Proline
			Methionine
			Glycine

[a] From (2).

Glutenin makes up approximately 55–70% of the gluten complex. Glutenin molecules are larger than gliadin molecules because of the high number of disulfide bonds connecting subunits of the entire molecule. Molecular weight estimations for glutenin range from 100,000 to several million. Glutenin subunits vary in size, and there is some evidence that gluten with certain high molecular weight glutenin subunits performs better in the breadmaking process than gluten without them. The disulfide bonding occurs toward the end of the chains, so, in effect, the glutenin molecule is linear. The tertiary structure is thought to be one containing repetitive β-turns, which form a β-spiral structure. This type of structure is stabilized by hydrogen bonding and may explain the elastic nature of glutenin. When stress is applied, this stable conformation is disrupted, but it returns when the stress is absent (3).

The albumins and globulins of wheat have a significantly different amino acid composition than the gluten proteins have (Table 3-3). There is less glutamine and proline and more of the basic amino acids

TABLE 3-3. Proportions of the Major Protein Fractions of Flour and Their Amino Acid Compositions (g/16 g of N)[a]

			Soluble Proteins		Gluten Proteins		
	Wheat	Flour	Albumin	Globulin	Gliadin	Glutenin	Residue Proteins
Extracting solvent			Water	0.5*M* NaCl	70% ethanol	0.5*M* acetic acid	...
Proportion,[b] %	...	100	15	3	33	16	33
Tryptophan	1.5	1.5	1.1	1.1	0.7	2.2	2.3
Lysine	2.3	1.9	3.2	5.9	0.5	1.5	2.4
Histidine	2.0	1.9	2.0	2.6	1.6	1.7	1.8
NH_3	3.5	3.9	2.5	1.9	4.7	3.8	3.5
Arginine	4.0	3.1	5.1	8.3	1.9	3.0	3.2
Aspartic acid	4.7	3.7	5.8	7.0	1.9	2.7	4.2
Threonine	2.4	2.4	3.1	3.3	1.5	2.4	2.7
Serine	4.2	4.4	4.5	4.8	3.8	4.7	4.8
Glutamic acid	30.3	34.7	22.6	15.5	41.1	34.2	31.4
Proline	10.1	11.8	8.9	5.0	14.3	10.7	9.3
Glycine	3.8	3.4	3.6	4.9	1.5	4.2	5.0
Alanine	3.1	2.6	4.3	4.9	1.5	2.3	3.0
Cysteine	2.8	2.8	6.2	5.4	2.7	2.2	2.1
Valine	3.6	3.4	4.7	4.6	2.7	3.2	3.6
Methionine	1.2	1.3	1.8	1.7	1.0	1.3	1.3
Isoleucine	3.0	3.1	3.0	3.2	3.2	2.7	2.8
Leucine	6.3	6.6	6.8	6.8	6.1	6.2	6.8
Tyrosine	2.7	2.8	3.4	2.9	2.2	3.4	2.8
Phenylalanine	4.6	4.8	4.0	3.5	6.0	4.1	3.8

[a] From (4).

[b] These figures are approximate; actual figures vary considerably depending on extraction procedure and flour sample.

and cysteine. Albumins and globulins usually have more secondary structure as a result. Many albumins and globulins are *enzymes*, which have a very defined tertiary structure. The spatial orientation of amino acids at the catalytic site of an enzyme is critical to the activity of the enzyme with respect to the reaction it catalyzes. In general, molecules of albumins and globulins are smaller than those of the gluten proteins.

Enzymes—Proteins that function as catalysts.

Glucose—A six-carbon sugar that is the "building block" of amylose, amylopectin, and β-glucan.

Carbohydrates—Substances (e.g., sugars and polysaccharides) that generally conform to the molecular structure $C_xH_{2x}O_x$.

STARCH

Starch makes up about 63–72% of flour (14% moisture basis) depending on the amount of protein. Unlike protein, starch consists of only one type of building block. It is made up entirely of *glucose*, a common six-carbon sugar that is also called dextrose and corn sugar. Glucose contains equal numbers of oxygen atoms and carbon atoms and twice as many hydrogen atoms (i.e., CH_2O)—hence the name *carbohydrate*. The chemical formula for glucose is therefore $C_6H_{12}O_6$. Although it can exist as a linear molecule, it also closes to form a ring (Fig. 3-2), which is the form that exists in starch polymers. The bond that forms when the ring closes is between the first and fifth carbons.

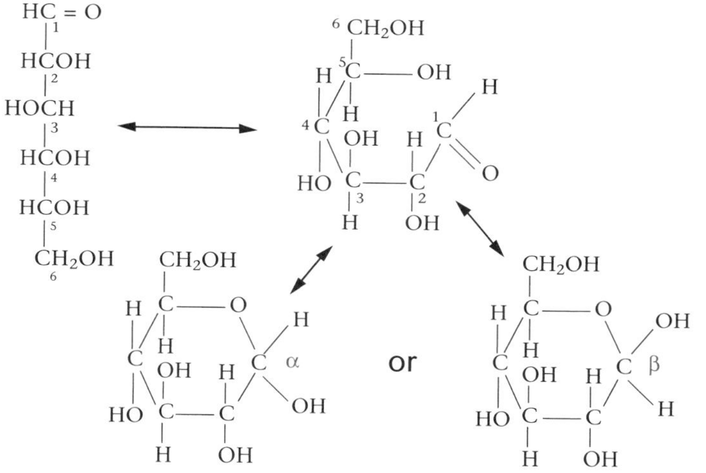

The glucose rings can be attached to each other through two bonds, the α-1,4 linkage and the α-1,6 linkage (Fig. 3-3). The number in this notation describes which carbons are linked. The α symbol denotes a spatial configuration in which the oxygen linking the carbons on adjacent glucose units is down. In the β configuration, it is up (Fig. 3-2). When these linkages are formed, water is lost. Hence the formula for the glucose units in a chain is $C_6H_{10}O_5$. The α-1,4 linkage is the primary bond connecting glucose units in starch, and it combines them so as to form a linear chain. In this chain, there is a "reducing end," which is the end where carbon 1 is not linked to another glucose unit. There is also a nonreducing end, which is the other end of the linear chain. Some glucose units will also

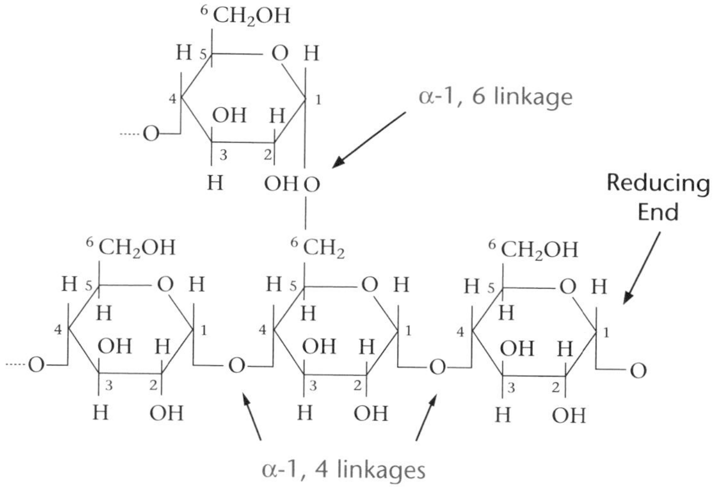

Fig. 3-2. Possible conformations for a molecule of glucose. (Reprinted from [5])

Fig. 3-3. α-1,4 and α-1,6 glycosidic linkages between glucose molecules in starch polymers. (Reprinted from [5])

be bound to another glucose unit through an α-1,6 linkage. This creates a branch point on the otherwise linear chain. When branch points occur in a starch molecule, more nonreducing ends are formed. There is still only one reducing end, however.

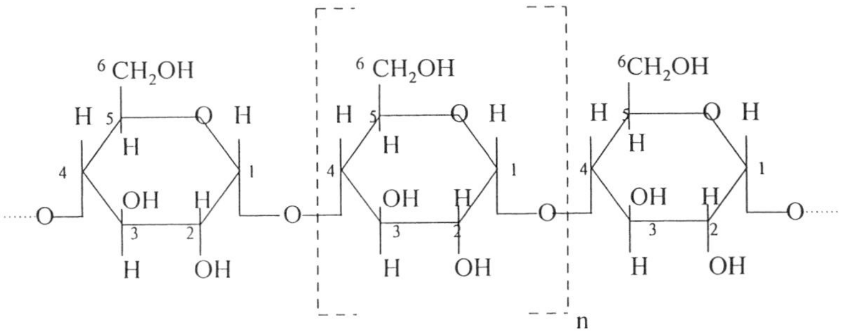

Fig. 3-4. Schematic diagrams of amylose structures. (Reprinted from [5])

Starch is made up of two different molecules of glucose. Amylose is an essentially linear α-1,4 linked molecule containing very few branch points (Fig. 3-4). There are usually between 1,500 and 6,000 glucose units in a single amylose molecule. Linear glucose chains, much like protein chains, tend to form helices. Thus, if amylose is allowed the freedom to assume any three-dimensional structure, as it is in solution, it will form a helix. This has important consequences for its properties and behavior. Common wheat starch is about 25% amylose.

The other molecule that makes up starch is amylopectin, which is distinguished from amylose in two ways. It is a highly branched and much larger. There are 300,000–3,000,000 glucose units in a single amylopectin molecule. Some (4–6%) of the linkages in amylopectin are α-1,6 linkages. The α-1,6 linkage inhibits helix formation, and therefore amylopectin has a less helical nature than amylose. Although an amylopectin molecule has many nonreducing ends, it has only one reducing end (Fig. 3-5).

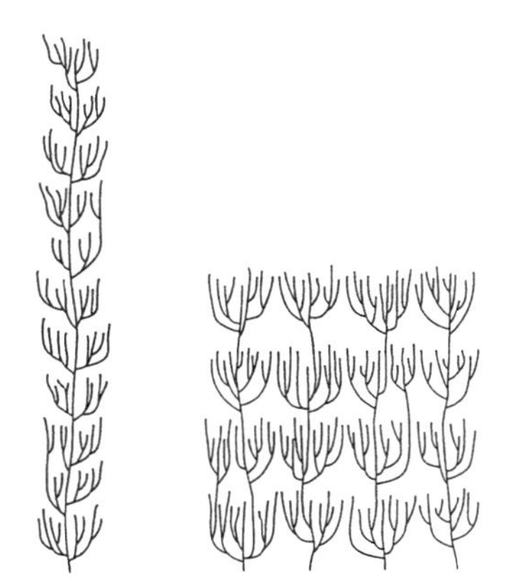

Fig. 3-5. Depiction of a portion of an amylopectin molecule (left) and an enlargement of typical packed clusters (right). (Reprinted from [6])

There is even a higher structure to starch. In its native form, the amylose and amylopectin molecules are contained within partially crystalline aggregates known as starch granules. These granules are about 30% crystalline. The amylopectin molecules within the granule provide the crystallinity, whereas the amylose is present in an amorphous form. Wheat starch granules are generally lens shaped and exist in two general size populations (Fig. 3-6). The smaller, more spherical granules (i.e., type B granules) are 1–3 μm in diameter, and the larger granules (i.e., type A granules) range from about 20 to 45 μm in diameter. Starch granules are highly ordered and can bend the plane of a polarized light. This property makes native starch granules *birefringent*. Bire-

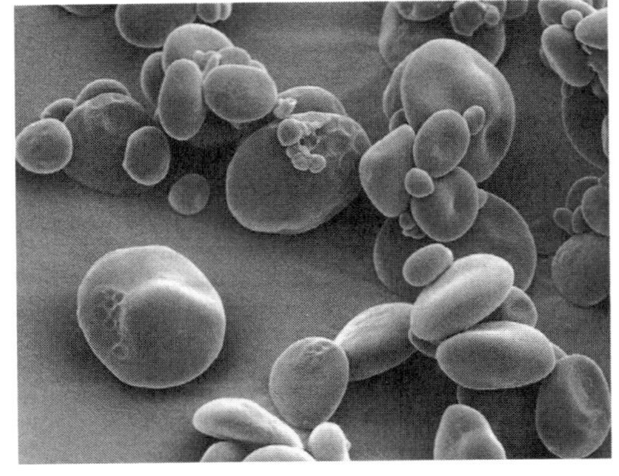

Fig. 3-6. Scanning electron micrograph of wheat starch. (Reprinted, with permission, from [7])

Birefringence—The phenomenon that occurs when polarized light interacts with a highly ordered structure, such as a crystal. A crossed diffraction pattern, often referred to as a "Maltese cross," is created by the rotation of polarized light by a crystal or highly ordered region, such as that found in starch granules.

fringent starch granules exhibit a "Maltese cross" pattern when viewed under a light microscope equipped with a polarizer (Fig. 3-7).

Starch structure and functionality are reviewed in detail in the AACC handbook *Starches* (5). The reader is referred to this reference for more information on this subject.

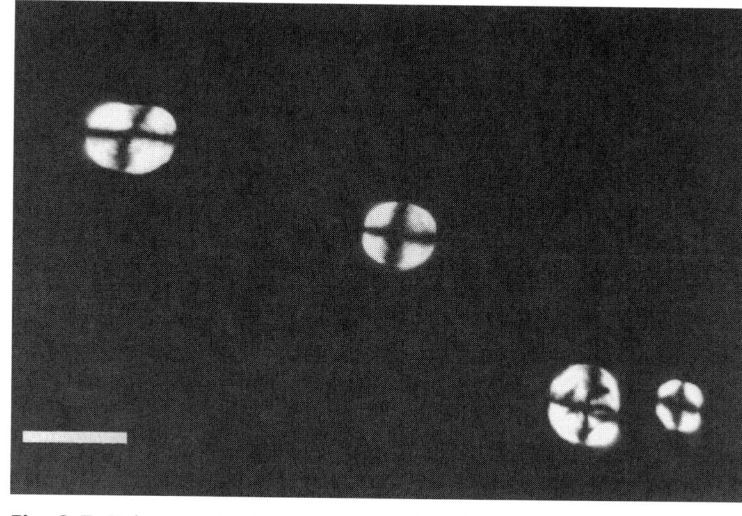

Fig. 3-7. Polarized light micrograph of native wheat starch. (Courtesy of Nancy Martin)

NONSTARCHY POLYSACCHARIDES

Flour contains other *polysaccharides* in addition to starch. A common name for some of these polymers is *hemicellulose*. However, this term applies to a variety of different types of polymers. Some of them are insoluble in water and others are highly soluble. The insoluble *pentosans*, which constitute about 2.4% of flour, are made up of the five-carbon sugars *arabinose* and *xylose*. The *arabinoxylans* are composed of chains of xylose units linked with α-1,4 bonds to side chains of a single arabinose unit, usually attached to carbon 3 of xylose (Fig. 3-8). Soluble arabinoxylans

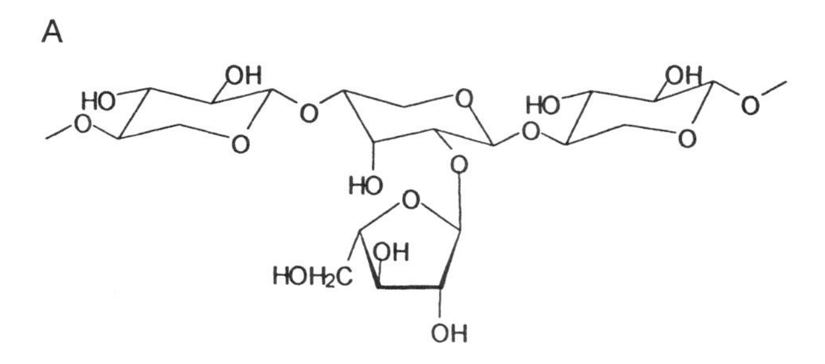

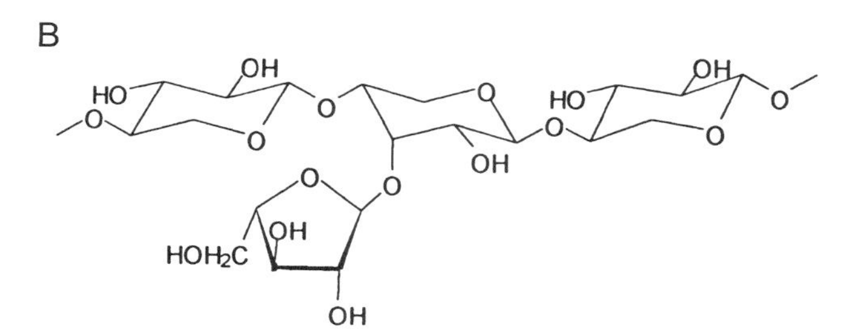

Fig. 3-8. Schematic diagram of the structure of arabinoxylan. In A, the arabinose unit is attached to carbon 2 of the xylose molecule; in B, it is attached to carbon 3. (Reprinted, with permission, from [8])

Polysaccharides—Polymers composed of sugar units.

Hemicellulose—A term often used interchangeably with "pentosan" to describe the nonstarchy polysaccharides of flour.

Pentosans—A group of polysaccharides containing five-carbon sugars and constituting a major portion of the nonstarchy polysaccharides of wheat.

Arabinose—A five-carbon sugar that composes the side chains of arabinoxylans.

Xylose—The five-carbon sugar that forms the backbone of arabinoxylans.

Arabinoxylan—A nonstarchy polysaccharide that is a major component of the cell walls of wheat.

β-Glucans—Nonstarchy polysaccharides composed entirely of glucose arranged linearly and bound together with β-1,3 and β-1,4 bonds.

Lipids—The class of compounds that includes fats and oils.

make up 1.0–1.5% of flour. Pentosan solubility is related to the molecular size and the number of arabinose side chains. The xylose "backbone" is very insoluble but becomes more soluble as more arabinose is bound. The molecule becomes less soluble as it increases in size.

β-Glucans are another type of nonstarchy polysaccharide found in wheat flour. The amount of β-glucans in flour is only about one-third the amount of the pentosans described above. β-Glucans are similar to starch in that they are composed solely of glucose. However, the linkages between the glucose units are not the same as those in starch. β-Glucans have both β-1,3 and β-1,4 linkages, and there are no branch points (Fig. 3-9).

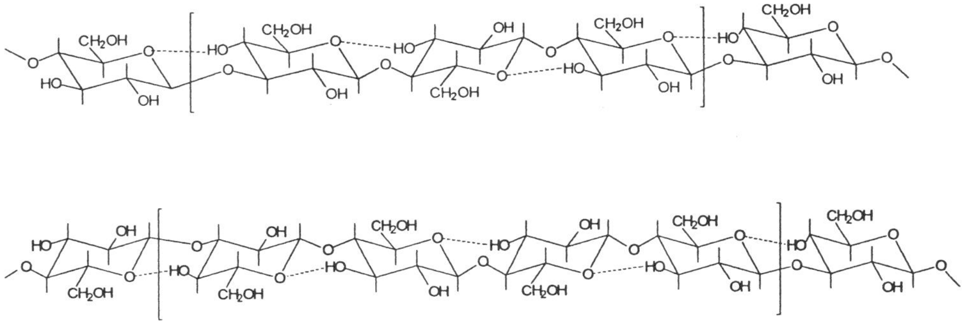

Fig. 3-9. Structure of mixed-linkage (1,3)(1,4)-β-glucan. The top row shows a β-1,3 unit, and the bottom is a β-1,4 unit. Both types of unit are found in β-glucan. (Reprinted, with permission, from [8])

Arabinoxylans and β-glucans are both natural components of the starchy endosperm, being concentrated in cell walls (Fig. 3-10). They are present in very large amounts in the aleurone layer. In some cases, arabinoxylans and β-glucans are considered contaminants of flour. Although present in low concentrations, they can affect the quality of the flour in significant ways because of their ability to bind large amounts of water.

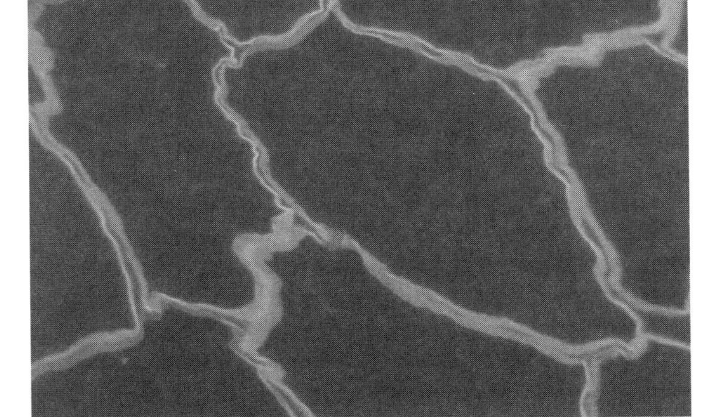

Fig. 3-10. Fluorescence micrograph of wheat endosperm stained with Calcofluor to show β-glucan. (Reprinted, with permission, from [8])

LIPIDS

The total *lipid* content of wheat endosperm is only about 1%. This isn't surprising. Most of the lipid comes from cell membranes, and, in the starchy endosperm, cells are large and cell membranes are very thin. The lipids are generally classified into free and bound lipids. Free lipids are extractable in a nonpolar solvent such as hexane, whereas bound lipids require more polar solvents such as ethanol for extraction. Another distinction between lipids is their polarity. *Triglycerides* and free fatty acids are considered nonpolar lipids. Lipids containing phosphorous groups (i.e., phospholipids) and those containing sugar groups (i.e., glycolipids) are considered polar lipids. Within the categories free and bound and polar and nonpolar, there are a large number of specific lipid structures. A couple of noteworthy flour lipids include the *tocopherols*, which have strong antioxidant qualities, and the *carotenoids*, which are the yellow pigments that give semolina (and, to a lesser extent, flour) its characteristic color.

Triglyceride—Three fatty acids bound to a glycerol molecule.

Tocopherol—Lipids found in flour that have antioxidant effects.

Carotenoids—The yellow pigments found in grains, including wheat.

Functionality of the Flour Components

Large companies that produce flour products may have hundreds of flour specifications, each required to make a specific product or meet a specific consumer need. Flour is a complex system, and because it is obtained from a plant, it contains the multitude of compounds found in any living tissue. These include the proteins, carbohydrates, and lipids discussed above as well as nucleic acids (e.g., DNA), vitamins, and minerals. Some of these components do not exert any influence on a product made from flour and are therefore nonfunctional. Other components play an important role in how a flour-based product behaves during processing or how well the final product meets the requirements of a consumer. The properties of these flour components are tested by the testing methods listed in flour specifications (Chapters 4 and 5).

GLUTEN

One highly functional component of flour is the gluten protein. The composition and subsequent functionality of gluten vary among and within wheat classes. For example, hard wheat (i.e., HRW or HRS) is considered good for breadmaking. The ability of flour made from these types of wheat to produce dough with good gas-holding properties is attributable to gluten. Soft wheat flour generally produces doughs with gas-holding properties inferior to those made from hard wheat flour. Therefore, in product applications such as breadmaking, in which a product is highly leavened and gas-holding ability is important, hard wheat gluten is generally more functional than soft wheat gluten. (However, flour made from other grains is radically inferior to flour from any type of wheat. Wheat gluten is unique with regard to its breadmaking ability.) In some products, though, soft wheat gluten has better functionality than hard wheat gluten. This is the case with crackers, cakes, and cookies, for which the gas-holding ability of the gluten before and during baking is not important. Hard wheat gluten in these products causes tough, unacceptable textures. Hence, gluten functionality is highly dependent on the product application.

The functionality of gluten is largely related to the physical properties of its components, glutenin and gliadin. When hydrated, gliadin is viscous and can be stretched to a thin strand or made to flow easily with gravity (Fig. 3-11). This property is called extensibility. Hydrated glutenin, however, is very elastic; there is considerable resistance when a mass of glutenin is stretched. If insufficient time is allowed for it to relax, it returns to its original shape. Combined, these two proteins yield the gluten complex, which is said to have *viscoelastic* (i.e., viscous and elastic) properties. When mixed in a dough with water and other components of flour, gluten forms a three-dimensional, continuous network. This provides the cohesiveness required to form a dough product. It also inhibits the diffusion of leavening gases though the dough when it is leavened and allows the product to rise before baking.

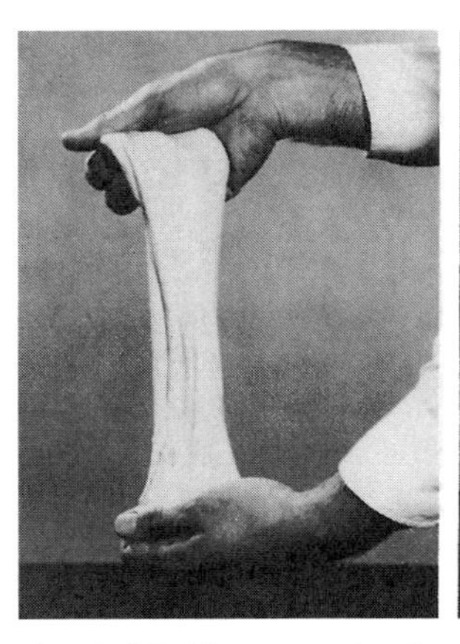

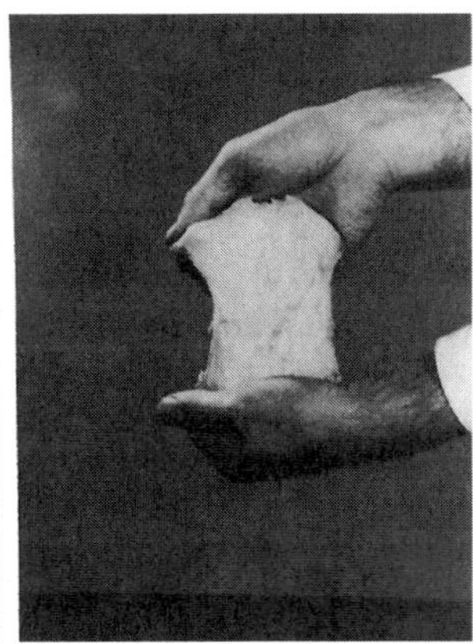

Fig. 3-11. Photographs demonstrating the extensibility of gluten (left) and its components gliadin (center) and glutenin (right). (Reprinted, with permission, from [1])

STARCH

Starch functionality is based largely on changes that occur when it is heated in the presence of water. The partially crystalline native granules lose crystallinity and birefringence during a process called *gelatinization*. In excess water, wheat starch gelatinizes at a temperature range of about 52–85°C. When heating is continued beyond this range, the starch granule starts to break down further, and eventually the granule is disrupted and amylose and amylopectin are dispersed. This process is called *pasting*. On cooling, the starch molecules can reassociate and form new crystal associations. This process is called *retrogradation*.

When starch gelatinizes and pastes, it binds much more water than it does in its native state. It is this property that yields the functionality of starch in many product systems. For example, during bread baking, there is a net movement of water from the hydrated gluten to the gelatinized starch. As the gas bubbles within the dough expand and eventually rupture, yielding an air-continuous system, the starch gel surrounding these bubbles increases in viscosity. This aids in the

Viscoelastic—Describing a substance that has both viscous (i.e., flow) and elastic properties.

Gelatinization—Irreversible loss of the molecular order of starch granules, shown by swelling and loss of crystallinity. Heating hydrated native wheat starch causes it to gelatinize.

Pasting—The breaking down of the starch granule following gelatinization.

Retrogradation—The process in which dispersed starch polymers recombine to form crystal structures.

formation of the final bread structure and greatly affects the crumb texture of the bread. During cake baking, the increase in the viscosity of the batter during the setting of the product structure is also governed by starch gelatinization in most cases. Starch gelatinization, pasting, and retrogradation play a major role in the formation of crusts on many baked products as well.

OTHER COMPONENTS

Water binders. The amount of water required to create correctly formulated dough or batter is dependent on all the water-binding components of the flour used. Because they occur in minor proportions relative to the amount of gluten and starch, these nonstarchy polysaccharides are often overlooked with respect to their functions. Two classes of nonstarchy polysaccharides, arabinoxylans and β-glucans, bind a great deal of water per unit weight. Up to one-third of the water in dough may be bound to these polymers. Damaged starch also plays a major role in determining the water relationships (Box 3-1) in flour-based doughs and batters. When granules are physically broken or sheared during milling, they bind much more water per unit weight than native starch does. Table 3-4 shows water uptake and distribution by some flour components.

Box 3-1. Water Relationships

Flours vary with respect to the amounts and, in some cases, the compositions of the gluten, starch, damaged starch, and nonstarchy polysaccharides that compose them. These are all hygroscopic components of flour, and they demonstrate their functionality only when hydrated. Therefore, the amount of water required to optimally formulate a dough or batter varies with the flour used. The water relationship, or the optimal flour-water combination for a batter or dough, must often be determined by trial and error, using the appropriate physical dough tests or batter viscosity tests.

TABLE 3-4. Water Uptake and Distribution in Dough[a]

		Water Uptake		
Constituent	Amount in 100 g Flour	(g/g, dry basis)	(g/100 g flour)	Water Distribution (%)
Starch (gelatinized)	68	...	...	...
Granular		0.44	25.4	26.4
Damaged		2.00	18.4	19.1
Proteins (gluten)	14	2.15	30.0	31.2
Pentosans	1.5	15	22.5	23.4

[a] From (9); used by permission.

Hydrolysis—A chemical process in which a peptide or glycosidic bond is severed after the addition of water.

Dextrins—Small starch fragments (linear or branched) formed when starch bonds are severed.

Lipids. The lipids natively present in flour can play a positive functional role in flour-based products. If cakes or breads are formulated with defatted flour, their volumes are depressed and textures are not optimal. In some porous dry products (e.g., breakfast cereals), flour lipids can break down to their component fatty acids (10) and then oxidize. This can lead to rancid odors and off-flavors, which are highly undesirable.

Flour Enzymes

AMYLASES

Amylases are enzymes that catalyze the *hydrolysis* of the bonds between glucose units in amylose and amylopectin. α-Amylase is an enzyme commonly found in flour that randomly attacks a starch polymer. It severs α-1,4 linkages anywhere in an amylose or amylopectin molecule, although it cannot hydrolyze α-1,6 linkages or any α-1,4 linkages that are in close proximity to a branch point. Consequently, if allowed to act long enough, α-amylase reduces starch to a series of branched and linear fragments (Fig. 3-12), called *dextrins*. The enzyme can act only very slowly on a native starch granule, but it can rapidly hydrolyze gelatinized starch. Although α-amylase denatures and becomes inactive upon heating, this flour enzyme is active above the gelatinization temperature of starch. Consequently, it can exert an effect on a final product (e.g., bread) despite a heating step (e.g., baking) during the process. α-Amylase is especially prevalent in flour that has been made from sprouted wheat.

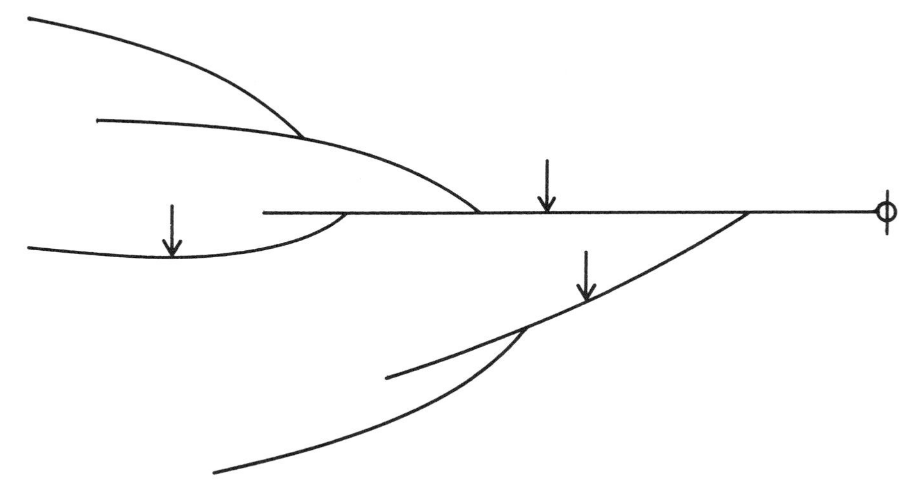

Fig. 3-12. Schematic diagram of the action of α-amylase on starch polymers. Arrows indicate points of attack on α-1,4-linked chains. (Reprinted from [2])

Flour also contains β-*amylase,* which also hydrolyzes α-1,4 linkages. The activity of this enzyme severs every second α-1,4 linkage, starting from the nonreducing ends of a starch polymer (Fig. 3-13), thus liberating the sugar *maltose*. Maltose is composed of two glucose units bound by an α-1,4 linkage. β-Amylase cannot sever α-1,6 linkages; hence, activity stops at a branch point. β-Amylase reduces amylose primarily to maltose, but it removes only the outer chains of amylopectin. The dextrin produced by the action of β-amylase on amylopectin is a called a β-limit dextrin. β-amylase cannot act at all on native starch. The β-amylase activity of flour is always high, even if its parent wheat was not sprouted.

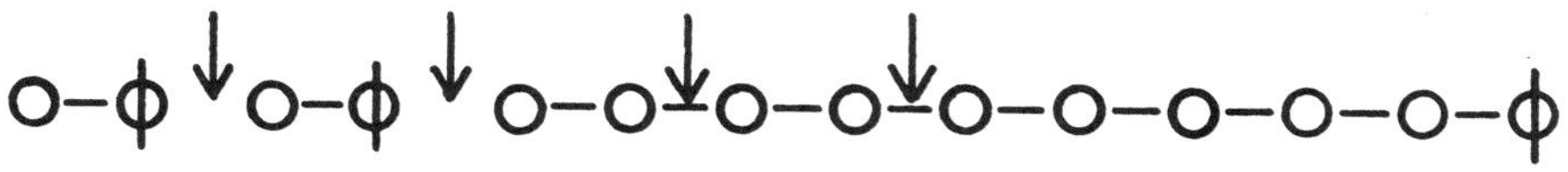

o–ϕ ↓ o–ϕ ↓ o–o ↓ o–o ↓ o–o–o–o–o–ϕ

Fig. 3-13. Schematic diagram of the action of β-amylase on starch polymers. Arrows indicate points of attack on α-1,4-linked chains. (Reprinted from [2])

The combined action of these two amylases reduces starch to maltose and small branched dextrins. Molecules such as these can enter into browning reactions; hence, when α-amylase activity is high in flour, finished products may show excess browning after baking or frying. Maltose can also be metabolized by yeast and therefore can be used to create leavening gases. The maltose produced in this manner augments the other sugars (e.g., sucrose and glucose) that may be added to a formula for this purpose. Additionally, when large starch polymers are reduced to smaller dextrins, the water-binding ability of the starch increases, and undesirable sticky or dense textures in the final product may result. This occurs because dextrins are more soluble than the large amylose and amylopectin molecules and therefore go into solution, producing a viscous syrup.

Although not generally added to batter or dough formulations, α-amylases from other sources are commercially available. The specificity of these enzymes is identical to that of wheat α-amylase, but significant differences in the temperatures of inactivation exist. α-Amylases from fungal sources lose activity in a range from ~65–75°C. Cereal amylases lose activity in a range from ~72 to ~85°C. Finally, α-amylases from bacterial sources are the most heat stable and lose activity in the range from ~82 to ~92°C.

PROTEASES

Proteases are enzymes that hydrolyze peptide bonds in proteins. Some proteases are present in flour, but their activity is generally low, and so they exert little effect on the gluten proteins or any of the other proteins in a flour-based product. Protease from bacteria or fungi can be added to dough to reduce the size of the gluten polymers. This

β-Amylase—An enzyme that hydrolyzes α-1,4 bonds in amylose and amylopectin.

Maltose—A sugar composed of two glucose molecules bound by α-1,4 bonds.

Proteases—Enzymes that catalyze the reaction that severs peptide bonds in proteins.

Lipase—An enzyme that catalyzes the reaction that severs bonds between fatty acids and glycerol in triglycerides.

Lipoxygenase—An enzyme that catalyzes the reaction between oxygen and double bonds in fatty acids and other lipids, often leading to rancidity.

Pentosanase—An enzyme that catalyzes the reaction that severs bonds in arabinoxylans.

Phytase—An enzyme that catalyzes the reaction that severs bonds between the phosphate groups and inositol in phytate.

Phytate—A molecule consisting of phosphate groups bound to the vitamin inositol.

makes dough easier to mix because the gluten polymers are smaller, but it may compromise the dough's gas-holding ability during fermentation and baking. Since enzymatic reactions can be difficult to control, chemical reagents such as sorbate, sodium metabisulfite, and cysteine hydrochloride, which interfere with the formation of disulfide bonds during mixing, are often added to achieve essentially the same effect on mixing.

LIPASES AND LIPOXYGENASES

Lipases are enzymes that hydrolyze bonds between fatty acids and the glycerol group in triglycerides. Because free fatty acids can impart a soapy flavor in a product system, their presence is not desirable. Additionally, when a fatty acid is liberated, the double bonds in it are very susceptible to the oxidation reactions leading to rancid off-flavors. Other flour enzymes, *lipoxygenases*, catalyze such reactions. Many of the carotenoid pigments of flour are also subject to the action of lipoxygenase. In most flour-based products, the destruction of these pigments and the resulting bleaching effect are positive. However, in pasta products, where the yellow color is desirable, this effect is detrimental, and semolina high in lipoxygenase activity is avoided.

PENTOSANASES

The pentosans of flour are susceptible to the action of enzymes that are collectively called *pentosanases*. Several specific enzymes hydrolyze pentosans. For example, one enzyme (xylanase) hydrolyzes the xylan backbone of the arabinoxylan molecule and another (arabinase) removes the arabinose side chains. Because pentosans bind a lot of water, pentosanases can affect the moisture relationships in a dough. This can cause significant problems in sensitive systems such as the doughs formulated to make crackers, frozen products, and refrigerated products.

PHYTASE

Phytase is a flour enzyme of nutritional significance. Much of the phosphate contained in wheat is bound in a molecule called *phytate*. The structure of phytate consists of phosphate groups bound to the six-carbon ring molecule inositol. Phytate is detrimental to the nutritional properties of wheat in a number of ways. Inositol is a vitamin, but neither the inositol nor the bound phosphate in phytate is available upon digestion in the human gut. Additionally, phytate chelates (binds) minerals such as calcium and magnesium, rendering them inaccessible to the consumer. The action of phytase, however, liberates the phosphate and increases the availability of all these nutrients.

POLYPHENOL OXIDASE

Polyphenol oxidase is an enzyme that catalyzes the polymerization of the phenolic compounds in flour. Phenolic compounds contain a ring structure similar to that of the organic compound benzene, and when oxygen is present, they polymerize to form very dark pigments. These pigments discolor dough and lead to a problem known as "gray dough," which is a surface phenomenon because oxygen is required for the reaction. Alleviating this problem requires the exclusion of oxygen or the inclusion of ascorbic acid (i.e., vitamin C) in the formula. Vitamin C inhibits polyphenol oxidase. It is also possible to limit the amount of polyphenol oxidase in the flour by using patent flours (see Chapter 2), in which much of the enzyme-containing aleurone has been removed.

Flour Additives

ENRICHMENT

In the 1930s, people in the United States were dying from the deficiency disease pellagra, which is caused by a lack of the vitamin niacin in the diet. In 1938, the government legislated that enrichment be added to bread flour to ensure that all people obtained enough of the B vitamins niacin, thiamine, and riboflavin in their diet (Table 3-5). Iron (i.e., usually ferrous sulfate or reduced iron) was also included in the enrichment at that time, and in the late 1990s, folic acid was included in enrichment standards as well (to reduce the incidence of a specific birth defect). The incidence of deaths from pellagra has radically declined as a result. Flour was chosen as the vehicle for these nutrients because it is present in foods consumed by essentially the entire population. Enrichment is usually added to flour at the mill, but it can also be included later in a product formula.

TABLE 3-5. Enrichment Requirements for Wheat Flour and Bread[a]

	Required Level (mg/lb)	
Enrichment	Flour	Bread
Thiamine	2.9	1.8
Riboflavin	1.8	1.1
Niacin	24.0	15.0
Iron	20.0	12.5
Folic acid	0.7	0.43
Calcium	960.0	600.0

[a] Source: 21 CFR 136, 139.

Enrichment in flour can cause some problems in flour-based products. Iron is a catalyst for the oxidative reactions leading to rancidity. It can also catalyze the polymerization of phenolics and contribute to the creation of gray dough. The form of iron used in the enrichment can affect these reactions. For example, ferrous sulfate is the most soluble iron form used in enrichments and the most available upon digestion. It is also the best catalyst for these reactions. Reduced iron is not as good a catalyst, but it is not assimilated as well.

Polyphenol oxidase—An enzyme that catalyzes a reaction between oxygen and the phenolic compounds in flour.

OXIDANTS

Oxidants are compounds added to dough formulations to change the characteristics of the gluten matrix (Table 3-6). These changes are commonly called "strengthening." Generally, this implies that the

Oxidants—Compounds added to dough to facilitate the formation of disulfide bonds.

TABLE 3-6. Oxidants and Reducing Agents in Doughs

	Maximum Level Permitted (ppm based on flour)
Oxidant	
Potassium bromate[a]	75
Potassium iodate[a]	75
Calcium bromate	75
Calcium iodate	75
Calcium peroxide	75
Azodicarbonamide	45
Ascorbic acid	GMP[b]
Reductants	
Potassium sorbate	GMP
L-Cysteine	90
Sodium metabisulfite	GMP

[a] Not allowed in the European Union.
[b] According to good manufacturing practices.

dough is made more elastic and its gas-holding ability is enhanced by facilitating the formation of disulfide bonds between glutenin subunits. Commonly used oxidants include azodicarbonamide, calcium peroxide, and ascorbic acid. Others approved for use include potassium iodate and calcium iodate. Potassium bromate was a very commonly used oxidant in bread products until recently, when it was categorized as a carcinogen. That status has affected its popularity and acceptance by consumers. Potassium bromate was unique because it was slow acting, and its strengthening effect occurred later in dough processing than those of other oxidants. Consequently, bromate allowed processing to occur before the dough strengthened and became difficult to handle. A significant effort has been made to find a replacement for bromate, but no universal replacer has been identified to date.

REDUCING AGENTS

Compounds used to inhibit the formation of disulfide bonds between gluten subunits and consequently make doughs mix faster and handle more easily are called *reducing agents*. Commonly used reducing agents include the amino acid cysteine, sodium metabisulfite, and potassium sorbate. These chemicals react with side groups of cysteine in the gluten protein and thus disallow their reaction with other cysteine side groups bound in other chains in the gluten. Reducing agents are effective in low concentrations (e.g., cysteine and sulfites up to 50 ppm, sorbate up to 500 ppm) and can both reduce mixing time significantly and make doughs easier to handle and process. Gas-holding properties are also affected, so reducing agents should be used cautiously with leavened products. Bromate is the only slow-acting oxidant that can reverse the effects of the fast-acting reducing agents on the gas-holding properties of dough. Therefore, because reducing agents and the fast-acting oxidants currently used have opposite effects, there is little sense in including both in a product formula. The best option is to source flour with appropriate processing and final product characteristics. If this is not possible, then the use of either a reducing agent or an oxidant should be considered, but the simultaneous use of both in a formula can never be rationalized.

Reducing agents—Chemicals added to dough to inhibit the formation of disulfide linkages.

CHLORINATION

Chlorination—The process of applying gaseous chlorine to flour.

Many soft wheat flours are treated with chlorine gas to improve cake-making properties. Chlorine is a very reactive molecule and reacts with many components of the flour. Although the mechanism of the improving action is not well understood, chlorine does affect the way water is bound by the flour and generally makes it more hydrophilic. Hence, batter viscosity of a chlorinated flour is generally higher than that of its unchlorinated counterpart. Chlorine destroys the pigments of flour; hence, chlorinated flours are very white. A byproduct of many *chlorination* reactions is hydrochloric acid. Because of this, the pH of chlorinated flours is lower than that of their untreated counterparts. This is used as a gauge of the degree of treatment. Bread flours are usually not chlorinated because gas-holding properties are compromised by chlorination. The "all purpose" flour sold in many supermarkets is generally a lightly chlorinated hard wheat flour. This type of flour can make both cakes and breads, but the quality is not high for either.

BLEACHING AGENTS

The most commonly used bleaching agent is benzoyl peroxide. It is slower-acting than chlorine and is added to flour as a powder, not a gas. It does not have an improving effect on soft wheat flours with respect to cake-making ability. Hard wheat flour is often treated with benzoyl peroxide when products with a white crumb are desirable. Calcium peroxide has a bleaching effect as well, but it also has an oxidizing effect on gluten.

MALT

Malt is ground, dried, germinated barley. During the production of malt, clean sound barley is first steeped to hydrate the grain and initiate the germination process. It is then allowed to germinate under cool, humid conditions for about five days and is finally kilned (i.e., dried) under controlled conditions. If high temperatures are employed during kilning, the enzymes in the system are partially inactivated, but a dark, rich malt with a characteristic "malt" flavor is obtained. Lower kilning temperatures produce malt with high enzyme activity but relatively low flavor impact.

Generally in a bread system, malt with high enzyme activity is used. The enzymes of primary importance are α-amylase and β-amylase because they can convert damaged starch in a bread dough to maltose. Maltose can be metabolized by yeast to produce carbon dioxide and ethanol, which are primary leavening gases in a bread system. Enzyme-active malt can also be added in high quantities to produce a sweet product formulated without sugar. Another major use of malt is in the production of breakfast cereals. Toasted malt is used in this application, primarily as a flavoring agent.

Nutritional Aspects

Wheat is generally a complete food, but some deficiencies should be noted. Gluten protein, for example, is low in some essential amino acids, most notably lysine. It is also true that milling "white" flour and removing the bran and aleurone reduces the vitamin, mineral, and fiber content. This has led many to believe that whole-wheat flour is "better for you" than white flour. When enrichment is added back, the vitamin content is clearly more than restored, but the fiber and mineral composition is not. Whether one favors white or whole-wheat, flour is a highly nutritious ingredient. Grain-based foods such as those made with flour form the base of the USDA nutritional pyramid, with the highest recommended servings per day of any of the food groups.

Web Site

American Association of Cereal Chemists—www.scisoc.org/aacc

The American Association of Cereal Chemists is an international organization of cereal science and other professionals studying the chemistry of cereal grains and their products or working in related fields. It publishes the journals *Cereal Chemistry* and *Cereal Foods World* as well as books on cereal and food science, including the textbook *Principles of Cereal Science and Technology*. The association holds an annual meeting, at which members present papers and make professional contacts. Member services include awards and honors, short courses on topics of interest to members, and a check sample service. Student memberships at reduced rates are also available.

References

1. Dimler, R. J. 1963. The key to wheat's utility. Baker's Dig. 37(1):52-57.
2. Hoseney, R. C. 1994. *Principles of Cereal Science and Technology*, 2nd ed. American Association of Cereal Chemists, St. Paul, MN. Chapter 4.
3. Tatham, A. S., Miflin, B. J., and Shewry, P. R. 1985. The beta-turn conformation in wheat gluten proteins: Relationship to elasticity. Cereal Chem. 62:405-412.
4. Wrigley, C. A., and Bietz, J. A. 1988. Proteins and amino acids. Pages 159-275 in: *Wheat: Chemistry and Technology*, 3rd ed., Vol. 1. Y. Pomeranz, Ed. American Association of Cereal Chemists, St. Paul, MN.
5. Thomas, D. J., and Atwell, W. A. 1999. *Starches*. American Association of Cereal Chemists, St. Paul, MN. Chapters 1 and 3.
6. Whistler, R. L., and BeMiller, J. N. 1997. *Carbohydrate Chemistry for Food Scientists*. American Association of Cereal Chemists, St. Paul, MN.
7. Evers, A. D. 1971. Scanning electron microscopy of wheat starch. III. Granule development in the endosperm. Starch/Staerke 23:157-162.
8. Hartunian Sowa, S. 1977 Nonstarch polysaccharides in wheat: Variation in structure and distribution. Ph.D. thesis, University of Minnesota, St. Paul, MN.

9. Bushuk, W. 1966. Distribution of water in dough and bread. Baker's Dig. 40:37-40.
10. Stauffer, C. E. 1996. *Fats and Oils*. American Association of Cereal Chemists, St. Paul, MN. Chapter 1.

Supplemental Reading

1. Hoseney, R. C. 1994. *Principles of Cereal Science and Technology*, 2nd ed. American Association of Cereal Chemists, St. Paul, MN. Chapters 2 and 10.
2. Morrison, W. R. 1988. Lipids. Pages 373-439 in: *Wheat: Chemistry and Technology*, 3rd ed., Vol. 1. Y. Pomeranz, Ed. American Association of Cereal Chemists, St. Paul, MN.
3. Pyler, E. J. 1988. *Baking Science and Technology*, 3rd ed., Vol. 1. Sosland Publishing Co., Merriam, KS.

CHAPTER 4

Wheat and Flour Testing

There are hundreds of wheat and flour tests, and it is beyond the scope of this handbook to discuss each one in detail. This discussion is limited to the procedures most commonly used in the testing of wheat and in the documents written to specify flour for end-user needs. The methods are described in general terms, and the importance of each test with respect to effects on processing and product quality is emphasized to ensure that the application of the test is understood.

The American Association of Cereal Chemists (AACC) and the AOAC International (formerly Association of Official Analytical Chemists) maintain collections of approved methods that may be particularly useful to those involved in the analysis of wheat-based products. These methods involve a detailed description of the purpose of each test, the apparatus and materials required, and a detailed procedure. Using an AACC or AOAC approved method enables laboratories to compare results and avoid discrepancies because identical, proven methods can be employed. These methods are also routinely upgraded to ensure that they are current with respect to analytical technology.

In This Chapter:

Tests Exclusive to Wheat
- Test Weight
- Thousand-Kernel Weight
- Kernel Hardness
- Flour Yield

Color Tests
- Pekar Color (Slick) Test
- Agtron Color Test

Odor Test

Basic Analyses
- Moisture
- Ash
- Protein
- pH
- Enrichment Detection
- Semolina Granulation

Flour Performance Tests
- Farinograph
- Mixograph
- Extensigraph
- Alveograph
- Gluten Washing Tests
- Alkaline Water Retention Capacity
- Solvent Retention Capacity Profile

Enzyme Analyses
- Falling Number
- Amylograph/Rapid Visco Analyser

Viscosity Methods

Microbial Assays

Baking Tests

Starch Tests
- Starch Content
- Starch Damage

Near-Infared Reflectance Methods

Tests Exclusive to Wheat

TEST WEIGHT

High test weight indicates sound wheat. As test weights drop, the percentage of small, malformed, and broken kernels usually increases. Hence, this test is used in the grading of wheat in many countries. Test weight of wheat is determined by weighing clean (i.e., dockage-free) wheat occupying a given volume. In the United States, test weight is reported as pounds per bushel, with the volume of a bushel defined as 2,150.42 in.^{3} or 35.24 L, while most other countries use the unit kilograms per hectoliter (as reported in AACC Method 55-10, the method for test weight). Test weights may range from about 45 lb/bu (57.9 kg/hl) for a poor wheat to about 64 lb/bu (82.4 kg/hl) for a sound wheat.

THOUSAND-KERNEL WEIGHT

Perhaps a better measurement of the soundness of wheat is the 1,000-kernel weight. This test simply measures the weight of 1,000 kernels of wheat. Because counting 1,000 kernels is a tedious process, equipment has been designed to do it automatically. HRS and HRW wheats range from about 20 to 32 g/1,000 kernels, whereas soft, white, and durum wheats range from 30 to 40 g/1,000 kernels. Generally, 1,000-kernel weight is reported on a 12% moisture basis.

KERNEL HARDNESS

Endosperm hardness can affect the amount of starch damaged during the milling process and subsequently the water requirements of the resulting flour. It varies significantly between wheat classes and even between varieties within a class. Growing conditions and moisture content of the kernel can also affect hardness. Measurements are based on near-infared absorption (AACC Method 39-70A), particle size of milled fractions (AACC Method 55-30), or on the force required to crush individual kernels (AACC Method 55-31).

FLOUR YIELD

This test is of importance to the miller because it is a general indication of how much flour can be made from a given wheat. It is conducted on a laboratory mill (e.g., Buhler or Miag Multomat) under a standard operating procedure. The value obtained (i.e., the percentage of flour based on the initial weight of wheat) relates to the extraction rate on a commercial mill. It should be understood, however, that a small laboratory mill is not as flexible as a commercial mill. Often, a miller can make adjustments in a commercial mill to make better separations and obtain yields higher than those obtainable on a laboratory mill.

Color Tests

PEKAR COLOR (SLICK) TEST

The Pekar Color Test (AACC Method 14-10) is a simple test, but it can be very useful to ensure that flour purity remains constant between flour shipments. For products in which bran specks are undesirable, this test can be especially valuable. Two or more 10- to 15-g samples of flour are placed side by side on a glass or metal plate. The flour is formed and smoothed ("slicked") to create a distinct boundary between the samples. Color differences and the number of bran specks can be easily compared. To accentuate differences, the flour samples can be sprayed with cold water and dried.

AGTRON COLOR TEST

A more sophisticated test for color, but one that is not widely used, involves the Agtron reflectance colorimeter (AACC Method 14-30). This instrument measures the reflectance of light from a flour-water slurry at a wavelength of 546 nm. The flour sample is calibrated against two disks of different reflectance to establish a range of brightness. Higher readings denote brighter color and generally purer flour. Straight-grade flour generally yields an Agtron color value of 80–85. Patent flours yield higher values, and high-ash flours yield lower values.

Odor Test

Flour is absorbent, and odors from various sources can be carried through to a final product. For example, flour stored near where a solvent has spilled will absorb the odor of the solvent. This odor and any associated flavors can carry through to the final product, even after it has been baked. Consequently, a very simple and valuable test often included in flour specifications is an odor test. A flour sample is simply evaluated for any unusual or undesired odors by someone in the quality lab. To facilitate the test, a standard flour can be maintained for comparison. Although this test is unsophisticated and subjective, it can be used to avoid major problems in final product quality.

Basic Analyses

MOISTURE

The complexity involved in the analysis of moisture is often underestimated. Moisture is easily driven out of a sample by heat when there is excess water in the system, but as a sample becomes drier, the remaining water can be tenaciously bound and, consequently, more difficult to remove. For this reason, it is very important to follow any moisture method exactly, especially with respect to temperature and time for oven-based methods.

The most commonly applied moisture-testing method for flour and wheat involves the use of an air oven (AACC Method 44-15A). Flour is analyzed directly, but wheat is ground in a laboratory mill before the analysis. If the sample is below 13% moisture, 2–3 g is weighed into a tared (preweighed) vessel. It is then heated for exactly 60 min at 130 ± 1°C, allowed to cool under desiccation, and weighed again. Percent moisture is calculated as the moisture loss divided by the original sample weight multiplied by 100. For samples over 13% moisture, an air-drying stage precedes this procedure (AACC Method 44-15A). Replicate determinations should agree within 0.2% moisture.

Moisture content over 14% affects the storage quality of flour and wheat. At higher moisture contents, mold growth, increases in microbial content, and infestation by insects are favored. High moisture can also lead to production problems because flour agglomerates more readily as it becomes wetter. This often causes hoppers and other devices with bottlenecks to "bridge" when flour clumps, resulting in blocked passageways.

Moisture should be included in every analysis of wheat or flour. It should also be the first test run, because it provides a basis for comparison for all other tests. Usually a 14% moisture basis is used for comparisons, but tests may also be compared on a dry basis or any other moisture basis. To assure accurate communication, it is important to always include the moisture basis when reporting wheat or flour analyses. Additionally, for other testing procedures where the dry matter in the sample governs the results (e.g., the farinograph test), it is important to know the moisture content of a sample to ensure that the appropriate amount of sample is analyzed. The following equation may be used to convert any analysis to any moisture basis:

$$A = B\left(\frac{100-C}{100-D}\right)$$

in which A = the analysis percentage at the desired moisture basis, B = the analysis percentage as originally analyzed, C = the desired moisture basis, and D = the moisture percentage as originally analyzed.

Example: A sample is analyzed and shown to contain 11.2% moisture and 12.5% protein. It is necessary to report the protein content on a 14% moisture basis.

$$12.1\%_{(14\%mb)} = 12.5\left(\frac{100-14}{100-11.2}\right)$$

ASH

The determination of ash (mineral) content requires that an accurately weighed sample be incinerated in a muffle furnace at or above 550°C until a constant weight is achieved (AACC Method 08-01). Ash content is calculated as the weight of the residue divided by the original weight of the sample, expressed as a percentage. As with most flour analysis results, it is usually reported on a 14% moisture basis.

Ash content is an extremely important measurement when it is necessary to know the purity of a flour sample with respect to bran contamination, because the residue remaining after this procedure represents the mineral content of the sample. Bran has approximately 30 times the ash content of the endosperm, so elevated test results reflect contamination with this highly concentrated ash source. Often, the ash test is used in conjunction with a visual test (e.g., AACC Method 14-10) to determine purity. In some cases (e.g., white wheat), ash content is a better indicator than a color test because the color of the bran is less apparent.

The ash test can also be important if minimizing the effects of an enzyme is necessary to ensure product quality. In cereals, enzymes are concentrated in the aleurone layer of the endosperm. Because the aleurone is closely associated with the bran, samples high in bran are also high in aleurone content. Hence, there is a good correlation between enzyme activity and bran content in many cases. It is often possible to reduce the enzyme activity of a flour by eliminating some high-ash streams (i.e., using a patent flour instead of a straight-grade flour).

PROTEIN

The primary procedure for determining protein content in wheat and flour has traditionally been the Kjeldahl method (AACC Method 46-10). There are several modifications of this basic procedure, but all are based on the same principle of converting protein nitrogen to ammonia, complexing it, and titrating it against a standardized sulfuric acid solution. Because the method measures nitrogen and not protein directly, the data must be converted. For wheat and wheat flour, the percent nitrogen obtained by the analysis is converted to percent protein by multiplying by a factor of 5.7. This factor varies for other proteins depending on amino acid composition.

Another method more commonly employed today is the Leco combustion method. This method (e.g., AACC Method 46-30) employs high-temperature pure-oxygen atmospheres to liberate nitrogen from the protein. Combustion methods are fully automated and are capable of measuring protein amounts in a range from 0.2 to 20.0%. Conversion from nitrogen to cereal protein by multiplying by 5.7 is also required for combustion procedures.

The most common spectroscopic means of determining nitrogen is near-infared (NIR) reflectance spectroscopy. This type of analysis has broad application and is widely used in mills to measure protein as well as many other parameters. NIR reflectance spectroscopy is discussed in a separate section later in this chapter.

Protein content is an important criterion for marketing and purchasing wheat and, as such, is included in almost every flour specification. In general, for hard wheats, the higher the protein content, the better the breadmaking characteristics of the flour. High-protein flour is also likely to require more water and mixing time to reach an optimum consistency for processing or product purposes. Bread doughs with high protein levels are generally more resistant to overworking during mixing. It is important to note, however, that for any processing or product parameter, protein quality (Box 4-1) plays at least as important a role as protein level. For example, high-protein flours with poor breadmaking qualities do exist, as do wheats with all combinations of protein level and processing characteristics.

pH

Testing the pH of a flour is required when it is important to know how much chlorine has been applied. The more chlorine applied, the lower the pH. Unchlorinated flour has a pH of 5.8–6.5, whereas a highly chlorinated flour can have a pH as low as 4.0. The test (AACC

Box 4-1. Protein Quality

Because protein quality isn't measured by content (amount) analyses, a second category of tests is commonly employed. Referred to as quality or performance tests, these are predictive in nature, seeking to predict the processing quality of an intermediate product (e.g., dough or batter) or the final quality of a food product (e.g., bread) from the properties of the flour.

Method 02-52), which is simple to perform, requires slurrying 10 g of flour in 100 ml of distilled water. The mixture is stirred for 15 min and allowed to settle, and the supernatant is decanted. The pH of the supernatant is measured with a pH meter.

ENRICHMENT DETECTION

To ensure that enrichment has been added to a flour, a test for the presence of at least one of the enrichment components must be employed. Perhaps the easiest test to perform is the visual inspection of the flour under ultraviolet irradiation. Bread enrichment particles glow when so irradiated. If a more specific test is required, approved methods are available for niacin (AACC Method 86-50A), riboflavin (AACC Method 86-70), thiamine, (AACC Method 86-80), and iron (AACC Method 40-40).

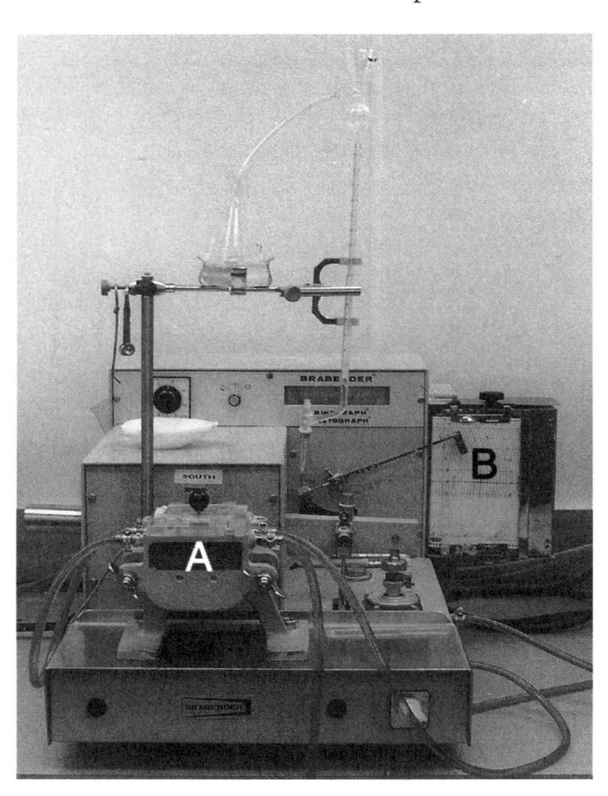

Fig. 4-1. The farinograph, with mixing bowl (A) and recording pen and chart (B).

SEMOLINA GRANULATION

The particle size of semolina is important to producing high-quality pasta. It is generally of a size such that no more than 10% will pass through a 180-µm sieve. If the particle size is too fine, the starch leaches into the cooking water and the pasta is mushy. AACC Method 66-20 describes a method for determining the particle size distribution of semolina, employing a series of sieves and a shaker. It can be adapted for flour and other granular materials, if required.

Flour Performance Tests

FARINOGRAPH

The farinograph is a type of recording mixer (Fig. 4-1). It measures and records the resistance offered over time by a dough against mixing blades operating at a constant speed (rpm) and a constant temperature. Parameters obtained from the resultant curve (i.e., resistance in Brabender units [BU] versus time in minutes) relate to the amount of water required to reach a desired peak consistency, the amount of time required to mix a dough to a desired consistency, and the amount of resistance a dough formulated with the flour will have to overmixing. Although the farinograph is but one of many recording dough mixers, it is the one that is most commonly used in the flour industry.

Farinograph absorption—The amount of water added to balance the farinograph curve on the 500-BU line, expressed as a percentage of the flour (14% mb).

Dough development time—A farinogram parameter, also called mixing time or peak time, which gives the time between the origin of the curve and its maximum.

For the test most widely used (AACC Method 54-21), the equivalent of 300 g of flour (14% mb) is placed in the farinograph bowl. The instrument is turned on, and water is added from a burette. As the flour hydrates and the dough forms, the resistance on the mixing blades increases, and the pen on the chart recorder and/or the curve on the computer screen rises. The mixing curve obtained generally rises to a maximum and then slowly falls from that point. To ensure that farinograms from different samples can be compared, the midpoint of the farinograph bandwidth at the maximum resistance is always centered on the 500-BU line. This is accomplished by adjusting the amount of flour and water used. An experienced operator can generally achieve this in the second or third run for any given flour sample.

A number of parameters can be derived from a farinograph curve (Fig. 4-2); the ones mentioned here are the most widely used to assess flour properties. The amount of water added to balance the curve on the 500-BU line, expressed as a percentage of the flour (14% mb), is the *farinograph absorption*. *Dough development time*, which is also called

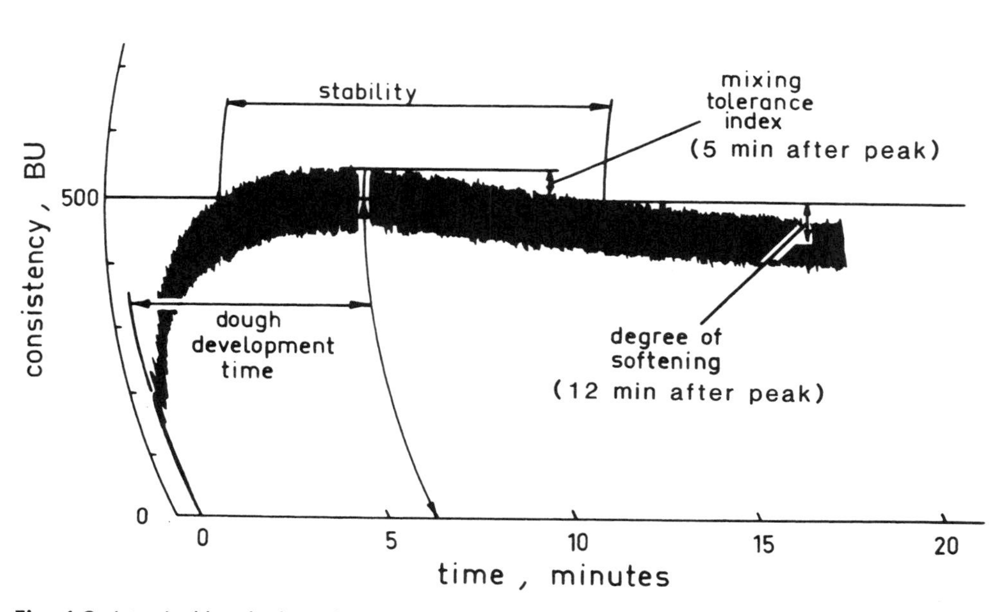

Fig. 4-2. A typical hard wheat farinogram, with some commonly measured indices indicated. (Reprinted from [1])

Tolerance index—A farinogram parameter(also called mixing tolerance index), measured as the difference in Brabender units between the top of the curve at the optimum and the point on the curve 5 min later.

Stability—A farinograph parameter, defined as the difference in minutes between the arrival time and the time at which the top of the curve falls below the 500-BU line (i.e., the departure time).

mixing time or peak time, is the time between the origin of the curve and its maximum. The maximum of the farinogram curve or any mixing curve is commonly believed to be the point at which the dough is optimally developed and best able to retain gas. Another parameter, called the arrival time, is the time between the origin and the point where the curve first reaches the 500-BU line. The *tolerance index* (also called mixing tolerance index, MTI) is measured as the difference in Brabender units between the top of the curve at the optimum and the point on the curve 5 min later. *Stability* is defined as the difference in minutes between the arrival time and the time the top of the curve falls below the 500 BU line (i.e., the departure time). C. W. Brabender Instruments, the manufacturer of the farinograph, has developed software to aid in the interpretation of farinograms. Generally, flour with good breadmaking characteristics has higher absorption, takes longer to mix, and is more tolerant to overmixing than poor-quality bread flour (Fig. 4-3).

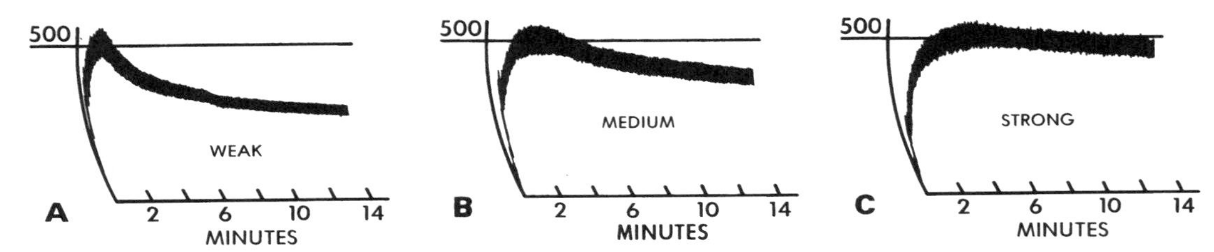

Fig. 4-3. Farinograms of three flours exhibiting weak (A), medium (B), and strong (C) mixing characteristics. The weak flour has a development time of 1.25 min and a tolerance index of 180. The strong flour has a development time of 5.0 min and a tolerance index of 30. (Reprinted from [2])

The parameters obtained from a farinogram are useful in determining the direction for adjustments required when flour changes, but they should not be taken as absolutes. For example, if the standard flour used in a commercial process has a farinograph mixing time of 4 min and a new shipment of this flour has a farinograph mixing time of 5 min, there is an indication that the mixing time in the commercial mixer where the new flour is used should be increased. However, it does not indicate that an increase of 1 min is required, because the mixers and ingredients in the mixers are different. Similarly, higher absorption indicates that more water is required to reach a desired consistency in a commercial process, but it doesn't indicate exactly how much. The mixing tolerance index and stability parameters indicate how well a flour resists additional work after it has been mixed to optimum. Long stability values and low tolerance index values are representative of flours that can be overmixed with little change in the consistency of the dough, but again, comparison of these parameters between flour shipments implies only directional changes. Sometimes useable relationships between farinograph parameters and commercial parameters can be developed, but it is important to ensure that the contributions of other ingredients in a formula and processing parameters such as temperature are also included.

As commonly used, the farinograph assesses flour properties through simple flour-water dough mixing. In addition to evaluating flour, the farinograph can be used to adjust flour-water ratios of fully formulated doughs in a commercial process. In this test, 480 g of dough is taken directly out of the commercial mixer at the end of the mixing cycle and put in the farinograph bowl, and a "remix" curve is run. The maximum of this curve can be used to specify the target consistency of a dough. Once the optimum consistency is identified, successive doughs can be evaluated and flour-water adjustments can be made to maintain this target consistency.

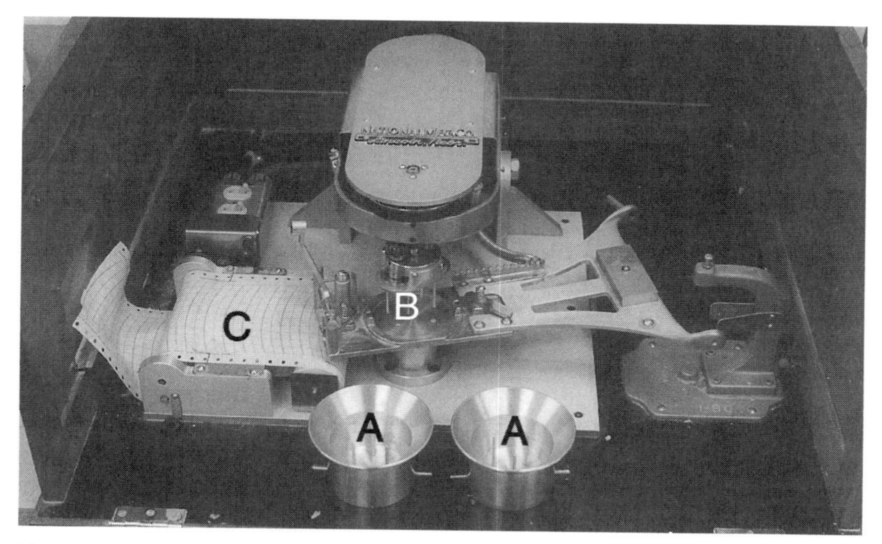

Fig. 4-4. A mixograph (10-g capacity) showing mixing bowls (A), mixing head and pins (B), and recording chart (C).

MIXOGRAPH

The mixograph (Fig. 4-4) is also a recording dough mixer, albeit a smaller, simpler piece of equipment than the farinograph. Samples as small as 2 g and as large as 35 g of flour can be evaluated (AACC Method 54-40A). Although the curve obtained from a mixograph (Fig. 4-5) also results from

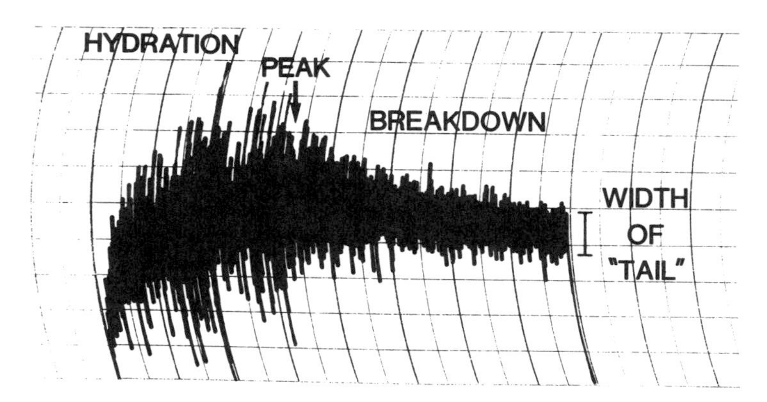

Fig. 4-5. A typical hard wheat flour mixogram, labeled to show the development (hydration), optimum (peak), and breakdown stages. (Reprinted from [3])

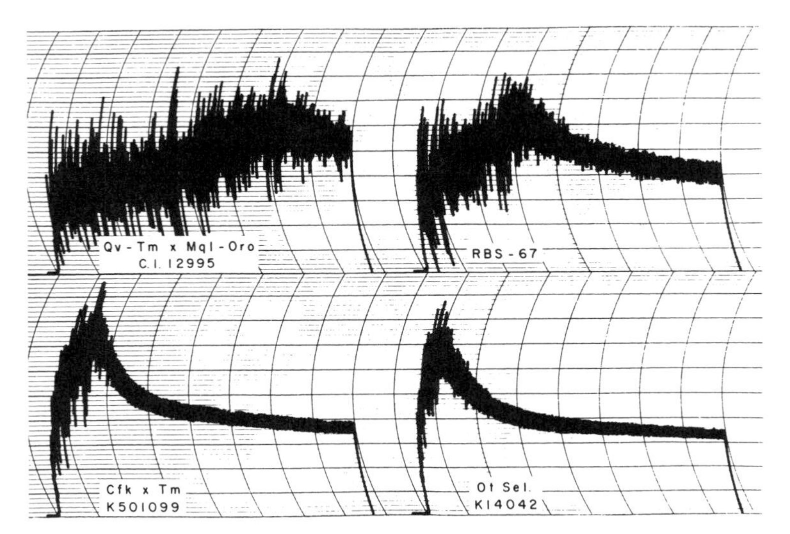

Fig. 4-6. Mixograms of hard winter wheat, showing strong (top left), good (top right), weak (bottom left), and extremely weak (bottom right) characteristics. (Reprinted, with permission, from [4])

the developing dough exerting force on mixing pins, the curve and the interpretation of it are not the same as for a farinogram. A "center curve" through the center of the mixogram band is first drawn to facilitate interpretation. All parameters are measured relative to this line. The height of the center curve at the highest point relates to the absorption as well as the protein content of the flour analyzed. The determination of optimum absorption (water addition) requires some experience and is best accomplished by comparing the test curve to the shape of curves with optimal absorption that have been produced previously. The time in minutes from the origin to the highest point on the center curve is the dough development time (also called time to maximum height, time to peak, and time to the point of minimum mobility). As on a farinogram, this is the point at which gas retention and ultimately bread quality will be the best. The range of stability is the time at which the mixogram band encompasses a line drawn parallel to the baseline at the highest point on the center curve. The shape of mixograph curves varies dramatically as the strength of the flour changes (Fig. 4-6). As with farinograms, software programs are available to analyze mixograms. Analogue and digital data collection, recording, and analysis are becoming increasingly common for both the farinograph and the mixograph. Plant breeders commonly use the mixograph because small samples can be run.

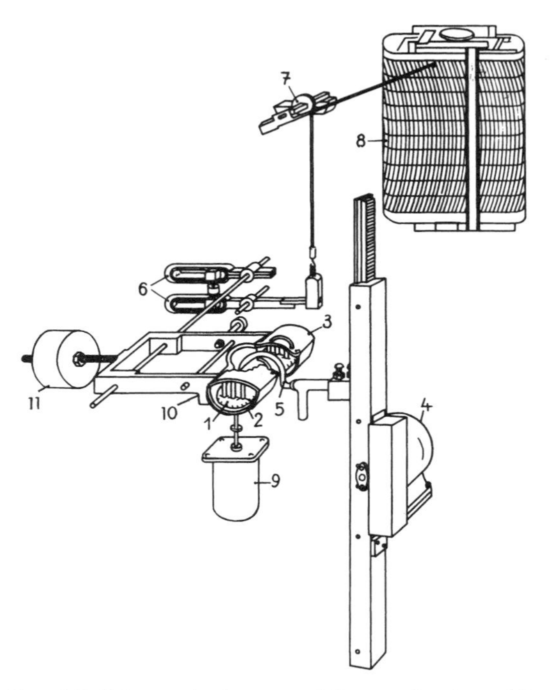

Fig. 4-7. Schematic diagram of the mechanism of the Brabender Extensigraph. 1 = test dough, 2 = cradle, 3 = clamp, 4 = motor, 5 = stretching hook, 6 = levers, 7 = scale head, 8 = recording chart, 9 = dashpot damper, 10 = arms of the balance system supporting the dough holder, 11 = counterweight of the balance system. (Courtesy Brabender Instrument Co.)

EXTENSIGRAPH

The extensigraph (Fig. 4-7) is designed to measure the extensibility and resistance to extension of a fully mixed, relaxed flour-water dough (AACC Method 54-10). A 150-g sample of dough is rounded and molded on the extensigraph and placed on a holder equipped with pins to ensure that the dough is secure. It is then allowed to rest for 45 min. After this time, a hook is drawn through the dough until it breaks. The curve obtained records resistance to extension (in BU) versus extensibility (in mm). Two parameters often derived from the curve are the height of the curve (i.e., resistance maximum) and the distance between the origin and the return of the curve to the baseline (extensibility). Figure 4-8 shows these commonly measured parameters. The area under the curve is also a useful parameter in applications in which both strength and extensibility are desired, such as in standard breadmaking processes.

The extensigraph is also often used to evaluate the effect of dough additives such as oxidants. After testing a flour-water dough containing an additive as described above, often the same dough will be reformed, relaxed for another period of time, and stretched

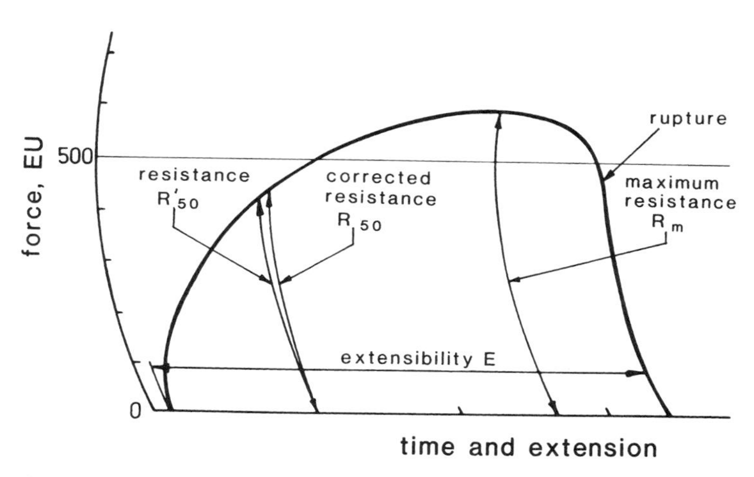

Fig. 4-8. Representative extensigram, showing some commonly measured indices. EU = extensigraph units. (Reprinted from [1])

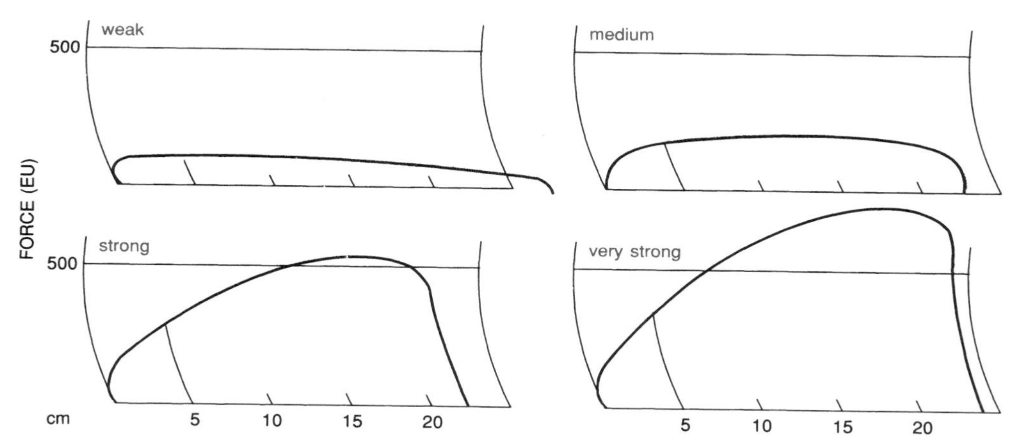

Fig. 4-9. Extensigrams of flours with weak, medium, strong, and very strong dough properties. EU = extensigraph units. (Reprinted from [5])

again to produce curves at another time. If repeated in this manner, the effect of the additive on resistance and extensibility can be observed with time (Fig. 4-9).

Perhaps the most common use of the extensigraph today is as a rough test of the viscoelastic properties of a formulated dough. A dough sample from commercial production is shaped, and the hook is drawn through it without any rest period. The resultant curve gives a rough estimation of the resistance and extensibility of the dough. It is important to control the time between sampling and evaluating the samples in this application of the extensigraph test because curves change as the dough is allowed to relax.

ALVEOGRAPH

The alveograph test involves mixing flour with a standardized salt solution, extruding the dough as a thin sheet, and expanding that dough sheet as a bubble (AACC Method 54-30A). The pressure inside the bubble is recorded with time until the bubble bursts. Usually, the test is repeated several times, and the parameters for the curves are averaged. The height of the peak relates to the resistance of the dough to deformation, while the length of the curve relates to its extensibility. The area under the curve represents the energy required to expand the dough and is related to the baking strength of the flour. This area is generally much larger for hard wheat flours than for soft wheat flours. Thus, the parameters obtained from an alveogram also describe the viscoelastic properties of a dough, but the measurement is distinctly different from those of the recording dough mixers or the extensigraph.

GLUTEN WASHING TESTS

Gluten washing tests such as AACC Methods 38-10 and 38-12 produce information concerning the quantity and quality of gluten in a flour or ground-wheat sample. These tests involve forming a dough and washing the starch and water-soluble components out of it. Wet gluten is left following the washing procedure, and the amount is an indication of both gluten quantity and quality. This is based on the observation that good-quality gluten binds more water than does poor-quality gluten. In AACC Method 38-10, the gluten is dried in an oven to determine the amount of dry gluten and hence the quantity. In AACC Method 38-12, which describes an automated procedure, a centrifugation step is included to dehydrate the gluten.

These tests are not used routinely in the United States, but they are employed extensively elsewhere. Wet gluten and dry gluten criteria are included in flour specifications in many countries as a primary test of flour quality. This is likely due to the simplicity of the test and the quantitative information obtained relating to both gluten content and quality.

ALKALINE WATER RETENTION CAPACITY

The alkaline water retention capacity (AWRC) is the amount of alkaline water retained by flour (14% mb) after a controlled centrifugation. Flour (1 g) is slurried with 5 ml of 0.1*N* sodium bicarbonate. It is then shaken, allowed to hydrate for 20 min, and centrifuged under specified and constant conditions of time and centrifugal force. The supernatant is decanted; the weight of the wet flour is determined; and AWRC is calculated. This parameter is important when the water relationships in a product are critical to product quality. One specific application of this test (AACC Method 56-10) is as a flour specification to predict cookie spread. As AWRC increases, cookie spread decreases.

SOLVENT RETENTION CAPACITY PROFILE

A more general test of the ability of a flour to retain solvents is described in AACC Method 56-11. The test is very similar to the alkaline water retention capacity test, but several solvents (i.e., water, 50% sucrose, 5% sodium carbonate, and 5% lactic acid) are used. The profile of solvent retention obtained can be useful for predicting the baking performance of a flour. In general, the higher the solvent retention, the better the baking quality. Consequently, this test may also be useful to include in flour specifications because it is an easy test to perform and yields information concerning the water requirements of the flour tested.

Enzyme Analyses

Flour contains many active enzymes. Most of them do not cause problems in processing flour-based products or in determining final product quality. The one notable exception is α-amylase, and so the methods discussed below are commonly required in flour specifications. Other enzyme tests are usually not specified. However, if a specific enzyme such as polyphenol oxidase or lipoxygenase is important, tests are available. A general reference on this subject (6) is listed at the end of this chapter.

FALLING NUMBER

The falling number procedure (AACC Method 56-81B) provides an index of the amount of α-amylase in a flour or ground-wheat sample. The procedure relies on the reduction in viscosity caused by the action of α-amylase on a starch paste and the fact that native wheat α-amylase is active above the gelatinization temperature of wheat starch. The procedure involves the use of the falling number apparatus (Fig. 4-10), which contains a boiling water bath, a stirring apparatus, a viscometer tube, and a timer. A flour slurry containing 7 g of flour and 25 g of water is added to the viscometer tube. The tube is immersed in the water bath and stirred. The starch gelatinizes, and the α-amylase liquefies the resultant paste. The time it takes (in seconds) for the viscometer stirring rod to fall through the starch paste is the *falling number*. Flour made from sprout-damaged wheat can have a falling number of 100 sec or lower. A bread wheat with average α-amylase activity has a falling number of about 250 sec. The upper limit for the falling number test is about 400 sec, which occurs for a flour devoid of α-amylase

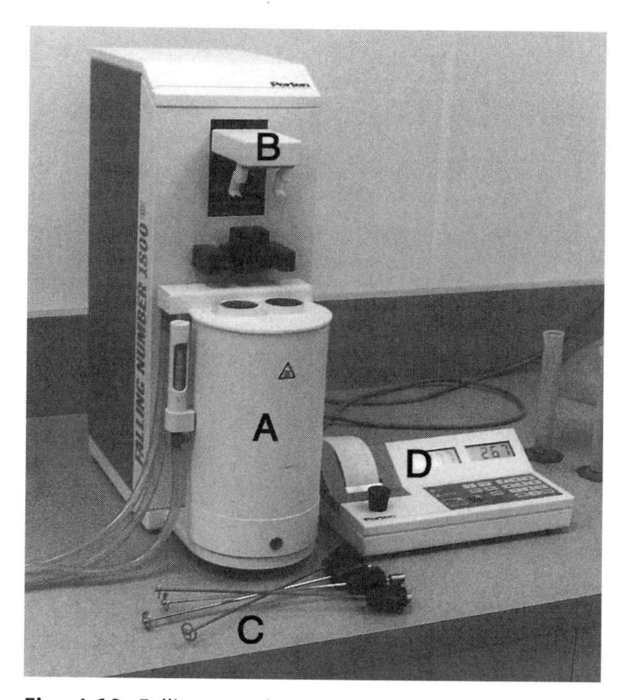

Fig. 4-10. Falling number apparatus, showing boiling water bath (A), stirring rod holder/actuator (B), stirring rod assemblies (C), and time/calculator (D).

Falling number—The time it takes (in seconds) for the viscometer stirring rod in a falling number apparatus to fall through the starch paste.

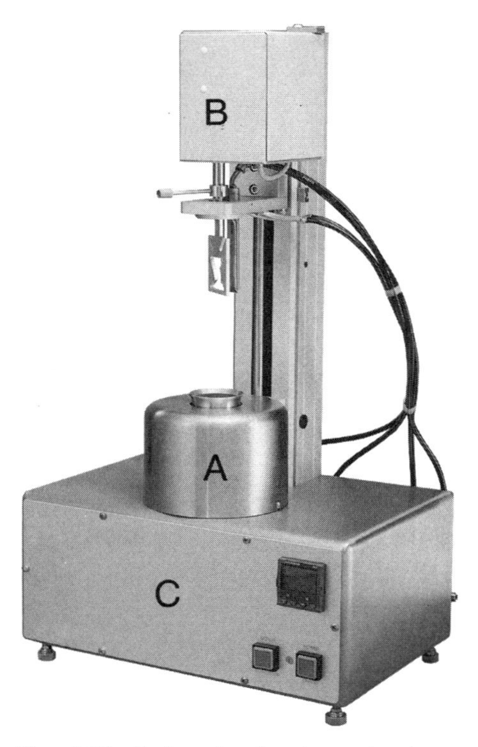

Fig. 4-11. Brabender Amylograph, showing sample holder and heating assembly (A), mixing head (B), and temperature and mixing control unit (C). (Courtesy Brabender Instruments Co.)

activity. The addition of malt or α-amylase from another source also affects falling numbers. Consequently, this test can be used as a means to monitor and control these additions.

Because α-amylase hydrolyzes starch linkages, more free sugars are liberated and lower starch paste viscosity results when enzyme activity is high. The implications of this for a baked product can be very significant, considering the functional roles that starch plays in most products. High α-amylase activity can lead to excessive browning since the reducing sugars liberated enter into Maillard browning reactions. Reduced viscosity caused by α-amylase can have devastating effects on batter products by reducing volume and producing an undesirable crumb structure. In bread products, the hydrolysis of the starch can lead to sticky crumbs and reduced volumes as well.

AMYLOGRAPH/RAPID VISCO ANALYSER

The effects of α-amylase on a flour-water system can also be evaluated with an amylograph (AACC Method 22-10) or a Rapid Visco Analyser (RVA, AACC Method 22-08). Figure 4-11 shows an amylograph and Figure 4-12 an RVA. Both of these instruments measure viscosity of a stirred flour-water slurry during a programmed temperature increase. As the temperature of the heating slurry rises, the starch gelatinizes and pastes and the viscosity rises. With the amylograph procedure, the recorded parameter is the maximum viscosity in

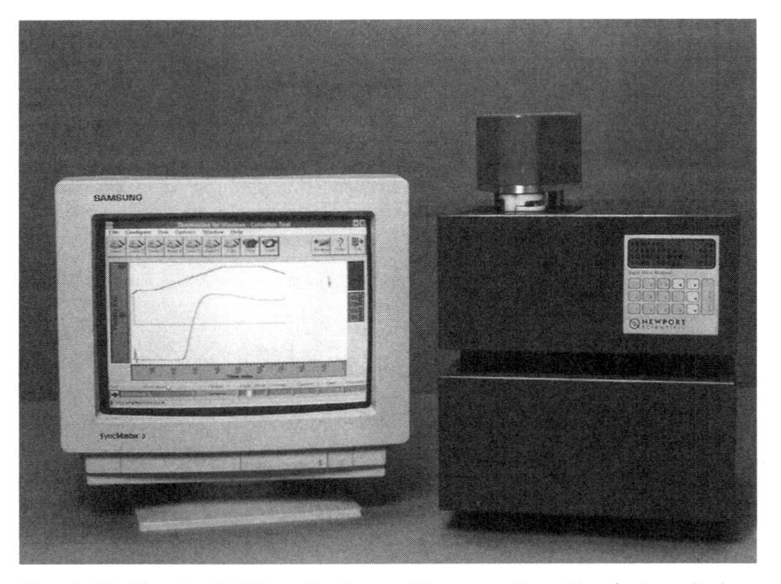

Fig. 4-12. The Rapid Visco Analyser. (Courtesy Foss North America)

Brabender units. Maximum viscosity declines with increasing levels of α-amylase (Table 4-1). With the RVA, the α-amylase activity is measured as the *stirring number* (SN). SN is defined as the apparent viscosity in RVA units after 3 min of stirring. As the enzyme activity rises, viscosity drops and SN increases. An RVA pasting curve is shown in Figure 4-13.

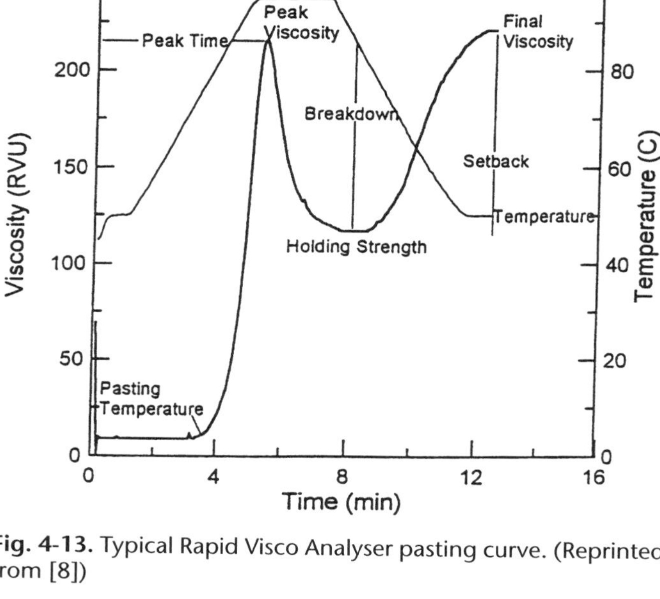

Fig. 4-13. Typical Rapid Visco Analyser pasting curve. (Reprinted from [8])

Viscosity Methods

Specifications for flours used in batter-based products often contain a viscosity test. Changes in the viscosity of a batter have dramatic effects on the volume and texture of a final product. The amylograph and the RVA can be used to define the entire pasting curve of a flour. As described in detail in *Starches* (9), the pasting curve includes an initial rise in viscosity as the starch granules gelatinize and begin to exude their components into the water. During this phase, the temperature of the stirred slurry is increased at a constant rate from 50 to 95°C. The viscosity rises to a peak and then declines as the temperature is held constant at 95°C and the suspension is stirred. This is the breakdown phase, which results from shear alignment of exuded starch components, which causes the viscosity drop. The temperature is then decreased at a constant rate, and the viscosity rises. This is the setback phase, in which the starch components begin to reassociate.

TABLE 4-1. Effect of α-Amylase on Viscosity Measured by Amylograph and Falling Number[a]

Malted Barley Flour, %	Amylograph (BU)	Falling Number (sec)
0.00	1,000+	415
0.05	970	320
0.10	570	273
0.25	360	199
0.40	255	176

[a] From (7).

Other types of viscometers (e.g., the Brookfield viscometer and the Haake viscometer) are useful as well. These instruments measure viscosity at a constant temperature, usually at or about room temperature. A viscosity test of this type is used to specify a flour for a specific batter product; hence, no broad standard procedure exists. Each test and the range of acceptable viscosity as it relates to product quality must be developed for each specific product. For example, assume that a test for the flour in a pancake formula has determined that the viscosity of a test batter made with a controlled amount of the desired flour and water is 180–200 cP. Using a specification for this flour that accepts only flour that yields viscosity in the range of 180–200 cP for controlled flour-water test batters may help maintain product quality.

Stirring number—The apparent viscosity in Rapid Visco Analyser units after 3 min of stirring.

Microbial Assays

Flour is a biologically active system and contains a number of microorganisms. Most flour-based products are baked, fried, and/or frozen, and therefore microbial contamination is usually not a major concern. However, in products that require no "kill" step, it may be important to include a test for one or more microorganisms. A general test that can be employed is the aerobic plate count (AACC Method 42-11). Specific tests also exist for coliform/*E. coli* (AACC Method 42-15), *Clostridium perfringens* (AACC Method 42-17A), *Salmonella* (AACC Method 42-25B), *Staphylococcus aureus* (AACC Method 42-30B), and other specific organisms or types of organisms.

Baking Tests

The best test for the quality of a flour as it relates to a baked product is to produce the baked product with the flour and measure the important quality criteria. In conducting a test of this type, it is very important to strictly control all the other ingredients in the formula and to ensure that none of the processing steps vary. This ensures that the observed effects are due to the flour. Of course, the test must be of a scale that can be performed before the flour is run on the commercial process, and it must also approximate the product and process as closely as possible. For these reasons, baking tests are often performed at the mill.

AACC baking quality methods exist for several standard types of baked products. The bread-baking quality of flour can be evaluated using straight-dough processing tests (AACC Method 10-09 and 10-10B) or a sponge-dough processing test (AACC Method 10-11). Flours for use in angel food cake and standard layer cakes can be evaluated with AACC Methods 10-15 and 10-90, respectively. A biscuit flour test (AACC Method 10-31B) and several cookie flour tests (AACC Methods 10-50D, 10-52, 10-53, and 10-54) have also been developed. These tests predict the quality of flour for use in products that are similar to the test product in formulation and processing. For other products, these tests should be adapted according to the specific formulation and/or processing requirements of the target product.

Starch Tests

STARCH CONTENT

Although starch is the major component of flour by weight, it is not often included in flour specifications. If required, AACC Methods 76-11 or 76-13 should be employed. These methods use starch-degrading enzymes after the starch in the sample has been gelatinized. The amount of glucose, the product of the enzyme action, is determined by a colorimetric assay.

STARCH DAMAGE

Variations in the amount of damaged starch (Box 4-2) in flour can cause changes in water requirements, dough mixing times, browning of crusts, and gas evolution in yeasted systems. Damaged starch is susceptible to hydrolysis by amylases, whereas native starch is not. AACC Method 76-31 exploits this difference by employing α-amylase to reduce the damaged starch to dextrins, followed by amyloglucosidase, which converts the dextrins to glucose. The glucose content is then determined by a spectophotometric assay. Damaged starch is expressed as a percentage of flour weight (as-is basis). Starch damage for HRS and HRW wheat flours is usually in the range of 8–12%. Soft wheat starch damage values are considerably lower, usually below 4%.

Box 4-2. Damaged Starch

Damaged starch refers to starch granules that have been physically altered from their native granular form. Starch granules broken in half are damaged. The halves still exhibit birefringence, are insoluble in cold water, and are not susceptible to enzymes. During milling, the shearing force can also produce damaged starch that does not exhibit these properties of native starch granules.

Near-Infared Reflectance Methods

NIR reflectance spectroscopy has been employed for many wheat and flour analytical determinations. Common NIR analyses include protein, ash, oil (fat), moisture, and kernel hardness. NIR analysis is based on computer analysis of the NIR reflectance spectra of wheat or flour samples and is especially valuable when large numbers of evaluations of the same material are required. To employ NIR analysis, it is first necessary to run a calibration set using known standards. For example, if a protein determination is desired, a set of samples with known Kjeldahl or Leco proteins must be used to calibrate the NIR instrument. For this reason, NIR is considered a secondary method, and the results can never be more accurate than the primary method used in the calibration. AACC methods exist for protein in flour (AACC Method 39-11), protein in wheat (AACC Method 39-25), and hardness of wheat (AACC Method 30-70A).

Web Sites

American Association of Cereal Chemists—www.scisoc.org/aacc

See description in Chapter 3.

AOAC International (formerly known as the Association of Official Analytical Chemists)—www.aoac.org

AOAC International is committed to the development and validation of methods in the analytical sciences and to the improvement of quality assurance procedures in laboratories.

References

1. Bloksma, A. H., and Bushuk, W. 1988. Rheology and chemistry of dough. Pages 131-217 in: *Wheat: Chemistry and Technology*, 3rd ed., Vol. 2. Y. Pomeranz, Ed. American Association of Cereal Chemists, St. Paul, MN.
2. Preston, K. R., and Kilborn, R. H. 1984. Pages 38-42 in: *The Farinograph Handbook*, 3rd ed. B. L. D'Appolonia and W. H. Kunerth, Eds. American Association of Cereal Chemists, St. Paul, MN.
3. Hoseney, R. C. 1994. *Principles of Cereal Science and Technology*, 2nd ed. American Association of Cereal Chemists, St. Paul, MN. Chapter 12.
4. Hoseney, R. C., and Finney, P. L. 1974. Mixing: A contrary view. Baker's Dig. 48:22.
5. Preston, K. R., and Hoseney, R. C. 1991. Applications of the extensigraph. Pages 13-19 in: *The Extensigraph Handbook*. V. F. Rasper and K. R. Preston, Eds. American Association of Cereal Chemists, St. Paul, MN.
6. Kruger, J. E., Lineback, D., and Stauffer, C. E., eds. 1987. *Enzymes and Their Role in Cereal Technology*. American Association of Cereal Chemists, St. Paul, MN.
7. Tipples, K. H. 1982. Uses and applications. Pages 12-24 in: *The Amylograph Handbook*, 2nd ed. W. C. Shuey and K. H. Tipples, Eds. American Association of Cereal Chemists, St. Paul, MN.
8. Zhou, M., Robards, K., Glennie-Homes, M., and Helliwell, S. 1998. Structure and pasting properties of oat starch. Cereal Chem. 75:273-281.
9. Thomas, D. J., and Atwell, W. A. 1999. *Starches*. American Association of Cereal Chemists, St. Paul, MN.

Supplemental Reading

1. American Association of Cereal Chemists. 2000. Approved Methods of the AACC, 10th ed. The Association, St. Paul, MN.
2. AOAC International. 2000. Official Methods of Analysis of AOAC International, 17th ed. The Association, Gaithersburg, MD.
3. D'Applonia, B. L., and Kunerth, W. H., Eds. 1984. *The Farinograph Handbook*, 3rd ed. American Association of Cereal Chemists, St. Paul, MN.
4. Faridi, H., and Rasper, V. F. 1987. *The Alveograph Handbook*. American Association of Cereal Chemists, St. Paul, MN.
5. Pyler, E. J. 1988. *Baking Science and Technology*, 3rd ed., Vol. 1. Sosland Publishing Co., Merriam, KS. Chapter 21.
6. Rasper, V. F., and Preston, K. R., Eds. 1991. *The Extensigraph Handbook*. American Association of Cereal Chemists, St. Paul, MN.

7. Shuey, W. C., and Tipples, K. H., Eds. 1982. *The Amylograph Handbook*, 2nd ed. American Association of Cereal Chemists, St. Paul, MN.
8. Williams, P., and Norris, K. 2001. *Near-Infrared Technology in the Agricultural and Food Industries*, 2nd ed. American Association of Cereal Chemists, St. Paul, MN.

CHAPTER 5

Specifying "Quality" Flour

In This Chapter:

Quality and Consistency
Communication
Flour Specifications
- General Information in a Comprehensive Flour Specification
- Testing Procedures
- Hard Wheat Products
- Soft Wheat Products
- Durum Products

Meeting and Enforcing Specifications
Crop Year Changeover
- Water Relationships and New-Crop Flour
- Gathering Information
- Assimilating the Information
- Disseminating the Information
- Upgrading Specifications

Specifications of any sort are often viewed as burdensome documents that do not accurately reflect the ingredient required to make the desired product and are, consequently, of little use. Writing specifications is generally not a task viewed with much enthusiasm. In fact, however, if written and managed well, specifications can be an invaluable means of assuring that product quality is high and consistent. Flour is the major ingredient in many products and consequently exerts a major effect on quality, however quality is defined. It is also a complex biological entity and, as such, varies significantly with the source of the wheat. For these reasons, it is important that flour specifications be given ample attention within any organization seeking to produce good-quality, flour-based products.

To be effective, flour specifications must be dynamic. It is unrealistic to believe that a flour specification will be accurate for more than a year, regardless of how much effort went into it initially. Wheat crops change in many ways from year to year, and the flour used in commercial processes may reflect these changes. Consequently, provisions should be made to upgrade flour specifications annually. Not all parts of a specification will require change every year, but it is likely that some parts will need modification. It is also important to keep a history of the changes made so that if compromises are required, the original specifications can be reinstated in later revisions.

Quality and Consistency

"Quality" is a subjective term that means different things to different people. For someone in a manufacturing setting, a high-quality flour might be one with processing characteristics that allow the machinery in the plant to run continuously without problems. A person in the quality assurance department of the same plant might consider a quality flour one that imparts a desirable characteristic to the finished product. For the purposes of the discussion herein, an average-quality flour is defined as flour that meets the expectations of the users, whoever they may be. Additionally, good-quality flour exceeds their expectations, and poor-quality flour does not meet their expectations.

Flour quality can vary, and often it is inconsistent flour that causes problems in a processing plant. For example, a poor-quality flour may be supplemented with an additive (e.g., vital wheat gluten) to achieve the performance required (e.g., better baked volume). If the flour doesn't change, the additive can be used continuously at the same level to achieve the desired result. If the flour changes frequently, however, frequent adjustments of the additive are required. Product not meeting standard will likely be produced because many changes will not be made until after a problem arises. Lack of flour consistency (i.e., batch to batch, day to day, shipment to shipment, and year to year) is the biggest issue facing processors of flour-based foods.

Communication

Problems with inconsistency can be minimized if there is good, honest, and frequent communication between the flour supplier and the flour user. If a mill cannot meet a requirement, it is important that the manufacturing plant receiving the flour be informed. Similarly, if the flour is within specification but not functioning properly, communication is important because the specification may not accurately describe the flour needed. Often, people experienced in dealing with flour-related problems can solve small problems and prevent them from becoming large ones if given adequate information concerning the problem. There are many ways to communicate, and whatever is effective should be used, but one primary aid for accurate communication is the flour specification. A concise, well-developed flour specification is critical to consistently supplying and receiving the flour required for production of a flour-based product.

The communication channel between the manufacturing plant and the supplying mill is very important, but other internal and external communication is important as well. A specification may be written for a flour that is used in several locations. If so, it is imperative that the people who write the specifications keep in touch with the people in the plants where the flour is used, to ensure that the specifications continue to meet their needs. Similarly, it is important for the people who buy the flour on a long-term basis to communicate with the milling company's management to ensure that factors that may affect the future quality and price of flour are clear to all.

Perhaps one of the most important avenues of communication is between the flour buyers and those involved in flour-related problem solving (e.g., quality assurance or research and development personnel). In many companies, this link is not good. Procuring flour at the lowest cost is an important goal of the flour buyer, but if the flour causes a problem in processing or end-product quality, the cost savings may be insignificant in comparison with losses incurred later in the chain. Hence, communication is critical to ensure that the appropriate flour, with respect to both cost and quality, is procured.

Flour Specifications

Specification systems vary among companies and come in many formats, but there is commonality with respect to the information that goes into a comprehensive flour specification. Although it is critical that the supplier have all the information necessary to meet the specification, it is also likely that parts of a specification may be considered confidential and not be shared with the supplier. The following section summarizes the general information that should be included in all flour specifications.

GENERAL INFORMATION IN A COMPREHENSIVE FLOUR SPECIFICATION

1. Designation. Usually the designation is numerical. Having a designation avoids confusion by ensuring that everyone involved with the flour refers to the same document.

2. Descriptive name. The name should be a concise description of the flour (for example, enriched HRW wheat flour). With computerized specification systems, this allows searches to identify similar flours.

3. Author. Specifying the author allows people to contact the right person with questions concerning the specification.

4. Comprehensive description. This section should fully describe the flour. Information should include wheat type(s) allowed in the blend; milling requirements (e.g., straight-grade, patent, stone-ground); additives or treatments (e.g., enrichments, malt, oxidants), with specified levels; and any other information that distinguishes this flour from others in the system.

5. Regulatory information. For most flour used in commercial processes in the United States, this section should state that the flour is required to comply with Federal Standard of Identity 21 CFR 137.165.

6. Food safety information. This section should include any information relevant to the safety of the flour for the consumer. Examples are statements concerning pesticide contamination, presence of toxins, and the allergenicity of wheat gluten (i.e., warnings to people with *celiac disease*).

7. Kosher designation. Not all flour is considered Kosher. For example, spring wheat in the United States is not always acceptable for use in Israel.

8. Sampling plan for supplier analyses. This section should describe the procedure for obtaining flour samples for analysis. It should include the frequency of analysis (e.g., per shipment, per mill run), how many samples are required, and information as to how the samples are physically collected.

Celiac disease—A chronic abdominal disease caused by allergy to one of the protein fractions of wheat.

9. Acceptance criteria. This section includes the testing procedures required for the supplier analyses. It should specify the approved procedures for the analyses and include the acceptable ranges for each test. (Later in this chapter, general testing procedures and those especially relevant for different types of flour and products are discussed.)

10. Supplier testing reporting requirements. The protocol for reporting test results is described here. Many companies require testing results to precede or accompany a flour shipment. Often these are included in a "certificate of analysis."

11. Receiving-plant testing requirements. Often a receiving plant tests each incoming flour shipment for only the most important criteria. Other tests may be on a more infrequent basis.

12. Procedure when a test is out of specification. When a flour tests within specification, it is accepted. This section should describe what action should be taken when an analysis is out of specification. This may involve informing the supplier, making adjustments in a formula, or, if severe enough, rejecting the flour.

13. Crop-year changeover procedures. Crop-year changeover can have significant effects on flour. Some companies prefer to evaluate blends of old-crop and new-crop flour through a changeover; others prefer to change abruptly. Regardless, all procedures concerning minimizing these changes and communicating any problems identified by the supplier or receiving manufacturing plant should be specified in this section.

14. Plant storage information. Temperature, humidity, and time criteria for storage in the plant should be located in this section. Generally, a plant does not keep more than a day or two of flour inventory, and consequently, long-term storage issues usually are not encountered.

15. Shipping information. Accurate communication on how the flour is to be transported to the plant is important. Common means are by railcar, flour truck, or individual bags. Size and weights of containers should also be specified.

16. Approved suppliers. Usually, a flour has one primary supplier because of a good working relationship between a mill and a plant. However, alternate suppliers who can supply the flour if the primary supplier cannot should also be identified.

17. Labeling information. This information states how the flour should appear on the ingredient declaration of the final product. This varies depending on the flour, additives, and the regulations of the country where the product is sold.

18. History of specification changes. A record should be kept of all changes made to the specification with time and the reasons for the changes. This enables changes that may compromise quality to be returned to previous standards when it is no longer necessary to compromise.

19. Confidential information. Some of the information above may be considered confidential by the company issuing the specification and should be kept in a section not distributed to the supplier. Additional information for a confidential section of the specification includes prioritization of the testing procedures. For example, a farinogram characteristic may be much more important than the protein level, and this may be critical information in situations where a flour shipment is out of specification. Other information included might be a list of the products made with the flour or explanations of the importance of critical testing procedures. The information in this section is certainly company-specific. It is obvious, however, that the portion of the specification shared with the supplier should have all the information necessary for supplying the flour required.

TESTING PROCEDURES

Tests for protein, moisture, ash, and odor and a visual test (e.g., Pekar slick test) are commonly found in all flour specifications. Another test that should be included in most flour specifications is one for amylase activity (e.g., falling number) because amylase activity can affect the appearance and texture of many flour-based products. These are the primary tests defining flour quality for many end-use characteristics. However, not all the tests discussed in Chapter 4 should be run on every flour. Some of the tests mentioned are appropriate only for certain types of flours and/or certain types of flour uses.

It is also important to note that all standard flour-testing procedures are somewhat general. They are designed to broadly describe an attribute of a flour that has an effect on many common products and processes. Because all products and manufacturing processes are different in some way, these standard tests do not always relate well to a specific end-use quality characteristic. *Therefore, it is very important to understand the relationships between a test parameter and a final product or processing characteristic.* In those cases where standard tests do not adequately relate to the important quality characteristics, it is necessary to adapt a standard test or develop a new specific test that better predicts them.

A test that would be very valuable in every flour specification is one in which the product is made on a small scale and the processing and product quality are evaluated (e.g., a baking test). However, these tests are usually not included in a flour specification for several reasons. First, to be highly predictive, the test must be somewhat product-specific, and millers would have great difficulty simulating the unique processing and formulation requirements of all their customers. Second, baking tests and other product-performance tests are complicated. They need to be well controlled, and the results can be influenced by many factors not related to the flour. Furthermore, they are also time-consuming, and any testing procedure is valuable only if the results can be obtained and reported quickly, certainly before the flour is used. Finally, the relationships between the specific final product parameters and the wheat or flour characteristics under the control of the miller are usually not well understood. Hence, millers may have no means at their disposal to make adjustments to meet the requirements of a product performance test.

HARD WHEAT PRODUCTS

Hard wheat flour is usually used in the production of yeast-leavened, dough-based products. Consequently, it is usually important to include at least one testing procedure that describes the water relationships and viscoelastic properties of a flour-water dough made with the flour. In the United States, a commonly used test is the farinograph test; in some other parts of the world, the alveograph test is the method preferred.

In yeasted products, the inclusion of malt may be specified. Unless it is heat treated, malt has a high amount of α-amylase, and any testing procedure to measure it is affected. A falling number, Rapid Visco Analyser (RVA), or amylograph test may be appropriate in such cases to ensure that the malt was added at the appropriate levels. Another test that may relate well to the performance of a flour in a yeasted system is starch damage. In addition to affecting water relationships in a dough, damaged starch can provide a substrate for yeast gas production and therefore may be important to measure.

SOFT WHEAT PRODUCTS

Many soft wheat products are batter-based. Consequently, viscosity measurements may be important and should be included. The amylograph or RVA may be useful for this purpose, but often a simpler test will suffice. Cold water viscosity is critical for many batter products such as pancakes and batters for fried coatings. For these types of tests, the type of viscometer, composition of the batter, and conditions used are specific to the product and some development of a testing procedure may be required.

For soft wheat flours that are chlorinated, it is important to include a pH measurement in the specification. When flour is chlorinated, its pH drops. Therefore, the pH is a good indicator of the extent of chlorination and perhaps of the functionality of the flour in the product.

Water requirements are very important to the production of some soft wheat products such as crackers and some types of cookies. A farinograph absorption test may be useful for this purpose. Solvent retention capacity tests may also be appropriate.

DURUM PRODUCTS

Color is an extremely important characteristic of durum products; therefore, an electronic color test may be included in a specification.

Specks in a pasta product are very undesirable, and a means of avoiding them (e.g., the Pekar slick test) definitely should be included. If color issues persist, it may be important to include a test for the enzyme lipoxygenase to ensure that its presence does not cause problems later in processing.

Another test that may be valuable is a granulation test. Semolina has a larger particle size than flour, and if a significant proportion of the particles are small, they can affect water relationships in processes (i.e., hydration rates) and can lead to some product quality issues such as sloughing (i.e., the breaking down of pasta product during preparation).

Meeting and Enforcing Specifications

All testing procedures are subject to variations, some associated with the sample and some associated with the testing procedure itself. Consequently, it is normal to observe fluctuations in the values obtained from analyses on flour samples that are all "within specification." It is important, therefore, to identify ranges of acceptability for any parameter obtained from a testing procedure.

To observe the amount of variation associated with a test, it is often useful to plot important test results from successive shipments. This type of graph is often called a "run chart." By drawing lines that correspond to the high and low values of the acceptable range, the frequency of "out of spec" analyses is easily observed. An occasional analysis close to, but outside of, the acceptable range is probably not reason for concern (Fig. 5-1). It may also be useful to determine the mean and standard deviations associated with the analyses and plot them as well. If a radical departure from the mean or a significant change in the standard deviation occurs (Fig. 5-2), it is likely time for some action to return the analyses to specified levels.

If several concurrent analyses are "out of spec," it is certainly time to contact the supplier, state that a change has been observed, and try to work together to return the test result to an acceptable range. If a comprehen-

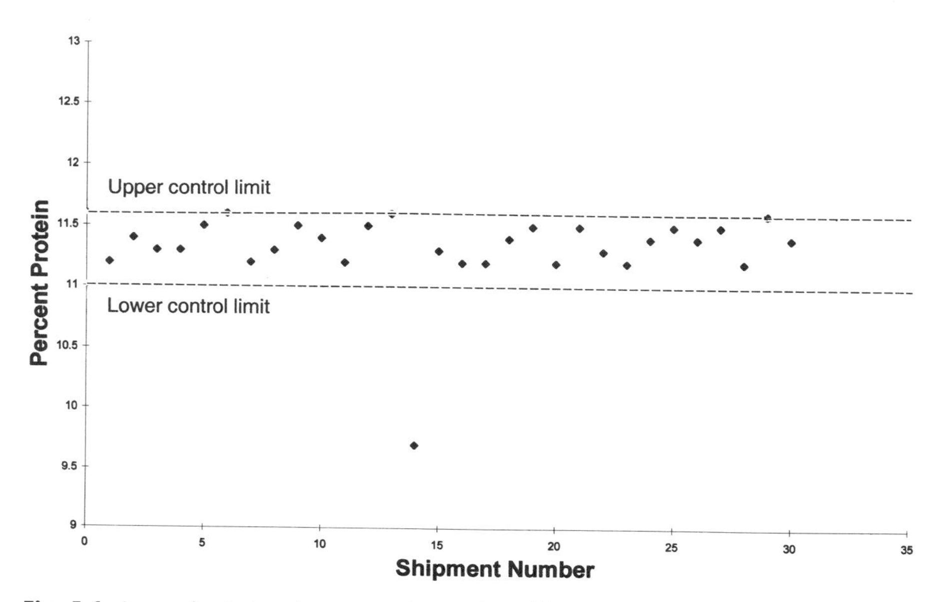

Fig. 5-1. A run chart showing percent protein, with one point varying from "normal."

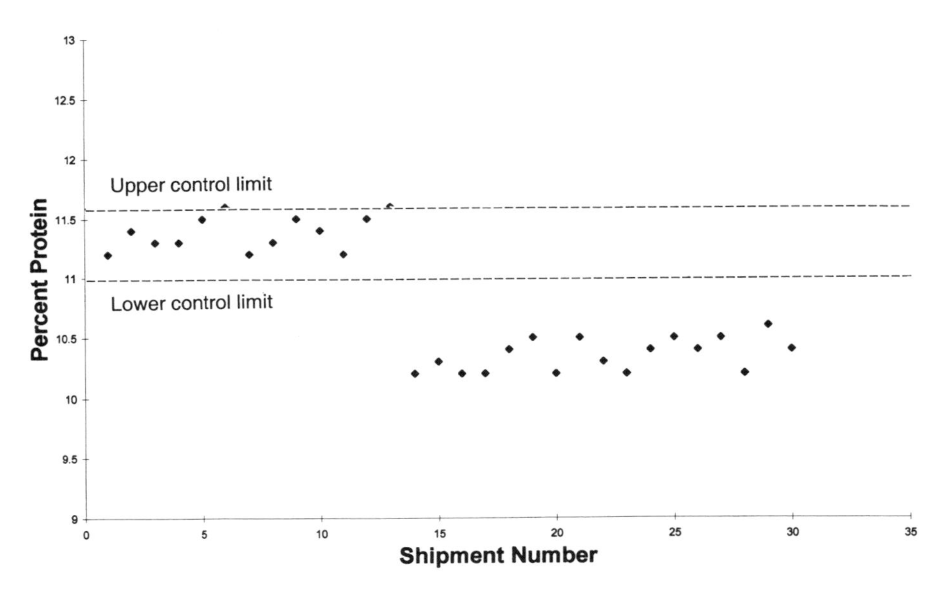

Fig. 5-2. A run chart showing percent protein, with a major departure from "normal."

sive, cooperative effort is made but does not achieve the desired results, it may be necessary to compromise and adjust the range of acceptability. If that is not a viable option, it may be necessary to obtain the flour from a different source. This is the primary reason that alternate suppliers should be identified for every flour used.

Crop-Year Changeover

Variation in flour quality due to a changing crop is the largest problem encountered by the processors of flour-based products. To ensure that crop-year changeover issues are minimized, it is prudent to gather as much information as possible concerning the new-crop wheat and to disseminate the important information to the appropriate people. Then, when the changeover occurs, some companies prefer to make the transition using blends of new-crop and old-crop flours and thus respond to the changeover over a period of time. Other companies prefer an abrupt change that allows them to address the problem once over a shorter time frame. After the issues with the new-crop flour have been resolved and the flour has been running consistently for some time, it is time to upgrade the specifications. This annual cycle of activities is necessary to avoid potentially very severe and costly crop-year changeover problems.

WATER RELATIONSHIPS AND NEW-CROP FLOUR

A phenomenon has been identified that affects the water relationships and hence the performance and variability of flour during some crop-year changes. This is the subject of U.S. patent 5,194,276 (1). When the water-binding capacity of new-crop flour is measured after milling, it rises with time and eventually reaches a stable plateau. Thus, the flour becomes more hydrophilic with time (Fig. 5-3). The time it takes for the water-binding capacity to stabilize changes with crop year. If the time is long, mill analyses and plant analyses do not agree because changes have occurred during the transport of the flour. As the wheat ages, the time it takes for the flour to stabilize is reduced, and the likelihood of disagreement between mill and plant tests decreases. A means of stabilizing new-crop wheat described in the patent involves raising the moisture content of the grain by about 5% and then rapidly drying it. This reportedly stabilizes the water-binding capacity of flour made from it so that there are no changes to yield test disagreements or fluctuations in processing or product quality as the flour ages.

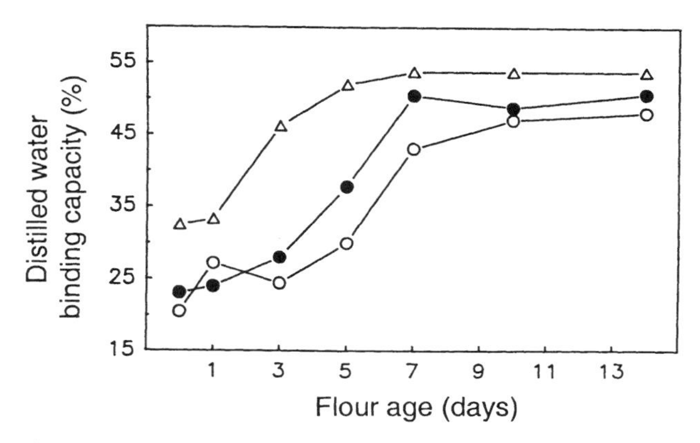

Fig. 5-3. Distilled water binding capacity (DWBC) vs. time after milling for wheats aged for increasing periods of time. ○ = zero weeks, ● = four weeks, △ = 12 weeks. (Reprinted from [2])

GATHERING INFORMATION

A tremendous volume of data is obtained every year on the developing crop of wheat, describing factors that ultimately affect the quantity and quality of the resultant new-crop flour. Information about soil conditions, weather data, area planted, growth progress, presence of diseases, etc. is routinely collected and reported on government Web sites (e.g., www.usda.gov/nass). Perhaps the best means of gaining information concerning an incoming crop of wheat is to attend a wheat crop tour. In the United States, these are organized by the Wheat Quality Council and involve a driving tour of each of the major wheat-growing regions. Participants estimate yields of developing wheat fields and obtain first-hand information concerning the new crop.

Data from the wheat tours provide information that primarily concerns the quantity of wheat that may be available at harvest. Some information concerning quality is obtained, but the primary means of obtaining preliminary quality information is through a crop survey. Most major milling companies conduct surveys of wheat-growing regions every year. These involve obtaining samples of the wheat immediately after it is harvested, milling it on a laboratory mill, and running standard tests to determine its general quality attributes. In the United States and many other countries, wheat-growing regions are divided into agricultural districts. For example, the state of Kansas is divided into nine districts. Consequently, composite samples of wheat from a district can be collected during a survey, and when data on the entire crop are reported, they can be segmented by district. This enables a wheat buyer to know the location of wheat with the appropriate quality attributes to meet the flour specifications of the mill's customers. Although the milling companies conduct most surveys, it is important for all involved in the production and use of the flour to have the information. The most advantageous relationships occur when mills and their customers share this information and work together to ensure that the appropriate wheat is purchased.

ASSIMILATING THE INFORMATION

The amount of information obtained on an incoming crop can be overwhelming, and much of it is not important to the end user. For this reason, it is advantageous to have a group of experienced people review the information, determine what is relevant to the specific flours used, and put the information into a usable form. There is no prescribed means of doing this, because it is specific to the requirements and resources available, but an example may be beneficial. In 1996, a drought occurred in the HRW region of the United States. Yield estimates obtained during crop tours were low. Because of the relationship between yield and protein level discussed in Chapter 1, the potential for high-protein wheat was clear. Protein level can affect many quality attributes, and this was clearly a piece of information that should have been reported to mills and to end users early in the crop-year changeover process. After surveys were completed, farinograph data indicated highly tolerant flour with long mixing times. These data were also significant and should also have been reported as soon as the information was available. Other information gathered on the 1996 HRW crop was within more normal ranges, but, clearly, protein level and strength, important factors affecting flour quality, were unusual and needed to be reported.

DISSEMINATING THE INFORMATION

The exact date of a flour change and any information indicating a change in a quality factor for the new-crop flour should be made available to a receiving manufacturing plant as soon as they are available and significantly before the flour arrives. Additionally, if an important quality factor is a matter of concern, it is prudent to have someone experienced in problems associated with it on site when the flour arrives. The communication channels discussed above are all important during a changeover, and the companies that communicate the right information at the earliest possible time will experience the fewest crop-year changeover problems.

Web Sites

United States Department of Agriculture, National Agricultural Statistics Service—www.usda.gov/nass

American agriculture is continually counted, measured, priced, analyzed, and reported to provide the facts needed by people working throughout this vast industry. Each year, the employees of USDA's National Agricultural Statistics Service (NASS) conduct hundreds of surveys and prepare reports covering virtually every facet of U.S. agriculture—production and supplies of food and fiber, prices paid and received by farmers, farm labor and wages, and farm aspects of the industry. In addition, NASS's 45 state statistical offices publish data about many of the same topics for local audiences.

Wheat Quality Council—www/wheatqualitycouncil.org

The main goal of the Wheat Quality Council is to enhance the milling and end-use qualities of all classes of wheat in all regions of the United States.

UPGRADING SPECIFICATIONS

When new-crop flour issues have been resolved and the flour is running in a plant consistently, it is time to review the specification and ensure that it accurately describes the new-crop flour. It is possible that one or more of the testing parameters requires adjustment or that something has changed in the materials or process used to produce the flour. The new-crop flour may be a quality improvement over the old-crop flour, and in that case, the new specification should be an improvement over the old one—a new standard has been developed. This positive change should be recorded in the specification so that it can be maintained. The opposite may also occur, however. If a compromise was required due to the new wheat, it is very important to record the changes in the new specification to enable the specification to be returned to the standard after a subsequent crop-year change. In all cases, the specification should meet the quality needs of the end user and allow the supplier sufficient latitude to meet the requirements. Consequently, people in the mill and in the receiving manufacturing plant should review every specification before approval.

References

1. Hoseney, R. C., Faubion, J. M., and Shelke, K. 1993. Method for rapidly producing stable flour from newly harvested wheat. U.S. patent 5,194,276.
2. Shelke, K., Hoseney, R. C., Faubion, J. M., and Curran, S. P. 1992. Age-related changes in the properties of batters made from flour milled from freshly harvested soft wheat. Cereal Chem. 69:145-147.

Supplemental Reading

1. Grant, E. L., and Leavenworth, R. S. 1988. *Statistical Quality Control,* 6th ed. McGraw Hill, New York.

CHAPTER 6

Products from Hard Wheat Flour: Problems, Causes, and Resolutions

Hard wheat flour is generally used to produce breads and related products. This is a very broad class of products, and the processes involved are as varied as the products. Besides flour, important ingredients in almost every such product are water, yeast, and salt. Shortening and sugar are also included in most formulas, as well as other ingredients like malt, preservatives, oxidants, and dough conditioners. The basic unit operations involved in making bread products are scaling, mixing, proofing (i.e., fermentation), forming, baking, and cooling.

In This Chapter:

Ingredients
- Breads
- Related Products

Processing
- Scaling
- Dough Processing
- Proofing
- Baking
- Cooling

Product Issues
- Appearance
- Texture
- Flavor
- Shelf Life Issues

Processing Issues

Fundamental Mechanism of Breadmaking

Troubleshooting

Ingredients

BREADS

Flour is the major ingredient in a bread formula (Table 6-1). Wheat flour, and more specifically the protein of wheat flour, is unique in its ability to form a dough that retains gas. This is a fundamental property required for the production of all leavened dough-based products. If gas is poorly retained, the volume of the product is low and the structure very dense. Yeast, which is included as the primary leavening agent, consumes sugars and produces carbon dioxide and ethanol. Hence, the yeast provides the gas to leaven the dough, to expand air cells, and ultimately to produce the cellular crumb structure of bread products. Salt is added to every bread formula as a flavoring agent, to strengthen the gluten, and to aid its ability to retain gas. If salt is not included in the formula, the product is bland, dense, and generally unpalatable. Alternatively, if too much salt is added, the

TABLE 6-1. Typical White Pan Bread Formula[a]

Ingredient	Percent (flour basis)
Bread flour	100
Water	63
Salt	2.25
Sugar	5
Nonfat milk solids	4
Shortening	3
Yeast	2
Yeast food	0.125

[a] Adapted from (1).

product has an unacceptable flavor and texture. Finally, water is required or a gas-retaining dough will not form. Water is a plasticizer and solvent for most of the ingredients. Without water, a gluten network cannot develop and the yeast is not able to produce sufficient gas to leaven the product.

Shortening is a common optional ingredient. It softens the crumb and extends shelf life by reducing firming. Like water, it is a plasticizer, so the proper balance of water and shortening must be determined in a formula to obtain the optimum viscoelastic properties of a dough. An additional benefit of adding at least a small amount of shortening is about a 10% increase in the volume of the bread.

Sugar and *yeast food* are other common optional ingredients. Although flour with a sufficient level of α-amylase contains enough sugar to maintain yeast fermentation for a significant time, additional sugar can extend fermentation times. Of course, sugar is also a flavoring agent that provides sweetness.

The addition of malt to a bread formula at an optimum level provides α-amylase and other enzymes beneficial to the process and the resulting product. The presence of these enzymes generally yields bread with better texture and reduces the occurrence of *keyholing*. Malt also is considered a flavoring agent. It can be provided in a heat-inactivated form, which imparts a toasted flavor to the bread.

Oxidants are present in most bread formulas to bolster specific volume. Legally approved oxidants in the United States include azodicarbonamide, calcium peroxide, potassium bromate, potassium and calcium iodate, and ascorbic acid (Table 6-2). They are effective in parts-per-million quantities. Potassium bromate, a very effective oxidant, is no longer widely used because it has been identified as a carcinogen, and unreacted bromate, albeit at very low levels, has been found in finished bread products. This has caused some complications in bread formulation because bromate is a slow-acting oxidant that is effective late in processing after significant work has been done on a dough. Identifying a replacement has been the subject of significant research, but, to date, an ingredient that replaces bromate in all applications has not been identified.

Dough conditioners (i.e., surfactants) are added to soften the crumb or to make the dough more resistant to work input during processing. Monoglycerides are the surfactants commonly used to soften the crumb. Dough strengtheners include sodium stearoyl lactylate (SSL), diacetyl tartaric acid esters of monoglycerides (DATEM), propylene glycol monostearate, sucrose monostearate, sorbitan monostearate, and ethoxylated monoglycerides.

Although all bread has a low bacterial load immediately after baking, airborne mold spores can contaminate bread after it has cooled. Hence, preservatives are often added to reduce the development of mold on the surface of bread products. The most commonly employed preservative is calcium propionate. Potassium sorbate is used

Yeast food—A minor ingredient composed of nutrients that enhance yeast activity. Prominent among these are ammonium salts.

Keyholing—Contraction and collapse of the side walls of a loaf of bread upon cooling.

TABLE 6-2. Additives Permitted by the U.S. Food and Drug Administration for Use in Baked Foods[a]

Additive	Function	Level Limit
L-Ascorbic acid	Oxidizing agent, dough conditioner	GMP[b]
Azodicarbonamide	Oxidizing agent	45 ppm[c] based on flour
Calcium bromate[d]	Oxidizing agent	75 ppm based on flour
Calcium diacetate	Leavening agent GMP	
Calcium iodate[d]	Oxidizing agent	75 ppm based on flour
Calcium peroxide[d]	Oxidizing agent	75 ppm based on flour
Calcium propionate	Mold inhibitor	GMP
Calcium stearoyl lactylate (CSL)	Surfactant, dough strengthener	0.5% based on flour
L-Cysteine	Mix time reducer	90 ppm based on flour
Diacetyl tartaric acid esters of mono- and diglycerides (DATEM)	Surfactant, dough strengthener	GMP
Ethoxylated mono- and diglycerides	Surfactant, dough strengthener	0.5% based on flour
Mono- and diglycerides	Surfactant, crumb softener	GMP
Monocalcium phosphate	Yeast food	0.25% based on flour
Polysorbate 60 in cakes[e]	Surfactant, crumb softener	0.46% of mix weight
Polysorbate 60 in bread[e]	Surfactant, crumb softener	0.5% based on flour
Polysorbate 65 in cakes[e]	Surfactant, crumb softener	0.32% of mix weight
Potassium bromate[d]	Oxidizing agent	75 ppm based on flour
Potassium iodate[d]	Oxidizing agent	75 ppm based on flour
Potassium sorbate	Mold inhibitor	GMP
Propylparaben	Mold inhibitor	GMP (0.1% max.)
Sodium benzoate	Mold inhibitor	GMP (0.1% max.)
Sodium metabisulfite	Mix time reducer	GMP
Sodium diacetate	Mold inhibitor	GMP
Sodium propionate	Mold inhibitor	GMP
Sodium stearoyl fumarate	Surfactant, dough conditioner	0.5% based on flour
Sodium stearoyl lactylate (SSL)	Surfactant, dough strengthener	0.5% based on flour
Sorbic acid	Mold inhibitor	GMP
Sorbitan monostearate in cakes[e]	Surfactant, emulsifier	0.61% of mix weight
Sorbitan monostearate in icing[e]	Surfactant, emulsifier	0.701% of batch weight
Succinylated monoglycerides	Surfactant, dough strengthener	0.5% based on flour

[a] From (1); used by permission.

[b] Good manufacturing practice. All the additives listed with a limit of GMP are substances generally recognized as safe (GRAS).

[c] Parts per million.

[d] If more than one of these oxidants are used in bread, the total weight used may not exceed 75 ppm based on flour, including any oxidants that have been added to the flour.

[e] If any combination of these additives is used, the limits on individual emulsifiers is lower than that listed here. Consult the appropriate paragraphs in the *Code of Federal Regulations* (21 CFR 172) for details.

sometimes in spray applications to reduce mold growth, but some consumers perceive an undesirable off-flavor.

It is always best to keep the number of ingredients in any formula to a minimum. One reason for this is that unnecessary ingredients can contribute significantly to product variation. Also, the process and final product quality are affected by the levels of all the functional ingredients in the formula. Product formulators have a tendency to try to identify new ingredients to solve problems, and so ingredient declarations tend to become longer with time. Each new ingredient holds the potential to interact with all the others in the system, and the functionality of ingredients added earlier may be compromised partially or completely. In some cases, simplifying a formula can solve a problem. This approach might be more beneficial than trying to solve a problem with a new ingredient, and it is always more cost effective.

RELATED PRODUCTS

Formulas for bagels, soft pretzels, and pizza crust contain a high amount of gluten. A high-protein (>13%) HRS wheat flour is the preferred base, and vital wheat gluten (up to 5%) is often added. This creates a finished product with the desired characteristic, a chewy bite. Dough-strengthening additives may also be employed. Very little, if any, shortening is included in these formulas.

Sweet doughs are those that contain more than about 12% sugar. Gas production by yeast is often inhibited because of the osmotic effects of the high sugar content on its metabolism. Consequently, it is common practice to increase yeast levels in sweet doughs as compared to formulas with less sugar. Other problems that may arise concern dough-handling properties. High-sugar doughs can be sticky, especially if they are warm. Therefore, high amounts of dusting flour during processing may be required. Additionally, the high quantity of sugar tends to dilute the gluten and yield a less cohesive dough. As a consequence, dough sheets may tear on high-speed sheeting lines.

Although many tortillas are corn-based, many others are made from wheat flour. A formula for wheat tortillas usually includes a high-protein flour, water, salt, chemical leavening agents, and shortening (about 8% of the total formula). Like bread, tortillas are susceptible to molds, and thus a preservative may also be included. Generally, chemical leavening is composed of an *acidulant* and sodium bicarbonate or (rarely) potassium bicarbonate. The acid produced by the acidulant causes the bicarbonate to produce gaseous carbon dioxide and consequently leaven the dough. In tortillas, the acidulants used are often sulfates (e.g., sodium aluminum sulfate [SAS], monocalcium sulfate, and calcium sulfate). Other acidulants commonly used are phosphate-based (e.g., monocalcium phosphate [MCP], dicalcium phosphate [DCP], sodium aluminum phosphate [SALP], sodium pyrophosphate [SAPP], and glucono-δ-lactone [GDL]).

Flat breads such as pita bread or pocket bread are made from a very simple formula, usually consisting of the four basic bread ingredients. The flour used in these types of bread is usually a very-high-extraction flour with high absorption. Although flat bread formulas do contain yeast, the distinctive void in the center of these breads is produced by steam leavening that results from the very high bake temperatures.

Acidulant—The acidic portion of a chemical leavening system.

Processing

SCALING

Scaling (i.e., weighing and delivering ingredients) is the first operation in the production of any product. Consistent and accurate ingredient weights delivered to the mixer are of paramount importance in attaining consistency and quality in a final product. Many industrial delivery systems vary considerably (i.e., ±5%) around a target weight. Minor ingredients are often premixed and added to the mixer with the major ingredients to minimize the variation of minor ingredients in a formula.

DOUGH PROCESSING

Straight-Dough Process. The straight-dough process is the simplest breadmaking procedure. It is used for hearth breads such as baguettes as well as for common pan breads. All ingredients are scaled and added to a mixer. The dough is mixed to optimum, which is often determined by interpreting a wattmeter curve. This is a curve of power versus time that resembles the curve of a recording dough mixer. The peak of the curve is the point of optimum development and the point at which the dough is best able to retain gas. After mixing, the dough is loaded into a trough and allowed to ferment in bulk for about 2 hr, during which the dough expands dramatically and the gas cells in it become larger. Large cells in the dough yield large cells and a very coarse structure in the final bread crumb. Therefore, to decrease the size and increase the number of these cells, the dough is "punched" or deflated one or more times during the bulk fermentation. Then the dough is divided into pieces of the appropriate size and weight, rounded into a spherical shape, sheeted, molded (i.e., rolled into a cylinder), and deposited in baking pans. It is then allowed to proof again in the pan, and when it reaches the desired height in the pan, it is baked.

The straight-dough process yields bread that is usually less flavorful than bread made from other processes because the fermentation times are shorter and flavor compounds are not produced in high quantities. Often, the bread can also be more elastic and chewy. Yeast liberates reducing compounds and proteases in doughs during long fermentations, and both of these reduce the size of the gluten polymers. In the straight-dough process, there is not sufficient time for this to occur, so the gluten polymers are larger, and the resultant crumb is more elastic. The short fermentation times used in straight-dough processing create some time inflexibility. Unlike sponge-dough processing, which allows some variation in fermentation time, straight-dough processes require carefully maintained production schedules, or viscoelastic properties will be significantly affected.

Sponge-Dough Process. A more popular process for breadmaking is the sponge-dough process. About two-thirds of the flour and water and all of the yeast are mixed, unloaded into a trough, and allowed to ferment in bulk. This is the sponge stage. Fermentation times for a sponge vary considerably throughout the industry but are usually 3–4.5 hr. The sponge and the rest of the ingredients are then added to the mixer, and the dough is mixed to its optimum point. This is the dough stage. The dough is allowed to relax for about 45 min and is then divided, rounded, sheeted, molded, and baked as in the straight-dough process. A commonly employed variation of the sponge-dough process involves a liquid preferment, in which all of the water and yeast and part of the flour are fermented in a tank before the dough is mixed.

Additional Dough Processing. The breadmaking processes described above are used for the pan and hearth breads commonly sold in retail outlets of various types (e.g., grocery stores, retail bakeries), and elements of these processes are used in other products as well. Various other products, however, require additional unit operations. Note that work input to the dough does not end in the mixer. Additional processing often imparts significant additional work, and a dough that was optimally mixed may be overmixed by the time processing is completed. For example, croissants, pastries, and other flaky products require lamination, in which dough layers are alternated with layers of butter or shortening. To laminate a dough, it is sheeted through a series of rollers; a layer of butter or shortening (i.e., "roll-in fat") is applied; the dough is refolded; and the sheeting and folding are repeated until the desired number of layers is obtained. There is a stretching action in any sheeting process, and turbulence is created in the dough as it is forced through the gap of each roller. The work imparted to the dough in this process is significant and may affect gas-holding properties. The turbulence may affect layer integrity. Consequently, "rest" time is often built into a process to allow some time for the dough to relax. This provides more process tolerance, but it also creates a more time-consuming process.

Many products are filled. In this case, the dough is often sheeted and the filling is deposited on the dough, after which the dough is folded and cut into the appropriate shape. Then the product is baked, frozen, or fried. Other products are filled after being baked by injection of the filling into the core of the product.

Immersion in boiling water is required to produce the chewy crust and smooth appearance characteristic of bagels. The boiling water gelatinizes the surface starch and creates the tough, chewy crust. In pretzel processing, lye is added to the water that serves as the boiling water bath. The high pH fosters Maillard browning, thus producing the shiny brown surface. Additionally, it produces the unique flavor notes characteristic of pretzels.

Mass-produced pizza crusts are generally sheeted and cut, or they may be pressed. A pizza dough is an elastic dough, however, and unless precautions are taken, the final pizza crust will not be round. To avoid this, cross rollers are often installed on the sheeting line to remove some of the directionality in the gluten network. If properly done, the "snap back" of the dough after cutting is the same in all directions, and the crust is round. Another pizza crust problem encountered in production is excessive blistering. Large blisters are undesirable. They are caused by large air pockets that form under the dough during sheeting, so the dough is often "docked" to reduce or eliminate the problem. Docking consists of running the dough under a roller with pins protruding from it. The dough is compressed at the points contacted by the pins, and the blisters that can form are smaller.

Tortillas can be produced by sheeting, dough pressing, or extrusion. The sheeting operation involves rolling rounded, proofed dough pieces, followed by hand stretching to the appropriate size and shape. The pressing operation, a more-automated procedure, involves pressing the rounded and proofed dough pieces in a heated, hydraulic stamping device. Finally, tortilla dough can also be extruded in a thin continuous band, followed by sheeting and cross rolling to the correct thickness and cutting with a circular die. Depending on the elasticity of the dough, the cutter may be designed to be slightly oblong to compensate for shrinkage after cutting.

Flat bread production is similar to the sheeting process for tortillas, except that the dough is usually sheeted mechanically and not as thin. The circular dough sheets are then proofed before baking.

PROOFING

Proofing is a rest period to allow the dough structure to expand from the action of the leavening. It can be intermediate in a process and followed by further dough manipulations, or it can be a final proof just before baking. The proofing process is critical to the volume, flavor, and texture of the final product.

In yeasted systems, carbon dioxide and ethanol are produced; both of these are leavening gases. Carbon dioxide production expands the dough during proofing and contributes to expansion during the early stages of baking. Ethanol leavens during baking only. Proofing also has effects on dough consistency and on the flavor of the final product. Sound (healthy) yeast has an oxidizing effect on the dough that changes its viscoelastic properties, rendering it more elastic and less extensible. Flavor precursors are also produced during yeast fermentation, giving bread its characteristic "yeasty" flavor.

Overproofing can lead to *lysis* of the yeast cells. When this occurs, the reducing agents and proteases contained in the yeast cells are liberated into the dough. A radical reducing effect can occur as a result, in which the dough becomes overly extensible or slack. Proteolysis of the gluten polymers can also lead to off-flavors, most notably "cheesy" or "cardboard-like" flavors in the final product.

Proofing is usually conducted in an environment with high temperature (i.e., 43°C, ~110°F) and high humidity (i.e., ~85% relative humidity). This enables the dough to expand but not to dry out and

Lysis—Rupture of the cell walls of a microorganism.

form undesirable crusts. Proofing can occur in bulk fermentations, as in the sponge-dough process, or in individual dough pieces, as in the final proof in a pan before bread baking. Equipment used in the proofing may be a humidity-controlled room for bulk fermentation or complex enclosed chambers with slowly circulating belts carrying trays of products.

BAKING

The baking process involves the application of heat and the conversion of the dough to a bread. During baking, a crust forms; the gas-retaining dough becomes an air-continuous crumb; flavor reactions take place in both the crust and crumb; and browning reactions occur in the crust. Many types of ovens are used in bread production, including reel ovens, tray ovens, and tunnel ovens. There are large differences in the construction of these ovens, but all are designed to provide uniform heat to all loaves in the oven. Pan-bread baking temperatures range from 190 to 232°C (375–450°F), and it usually takes about 15–25 min to bake a 1-lb (454 g) loaf, but there is considerable variation in bake temperature and time depending on the type of product. Some ovens are equipped to inject steam into the oven atmosphere. This provides more water on the surface of the dough, a more viscous starch gel, and a thicker final crust.

Generally, lean formulas (i.e., those low in fat and sugar) require higher bake temperatures and shorter bake times than do rich formulas. Considerable variation exists in optimal baking conditions, however. For example, flat breads are baked in a high-temperature oven (410–499°C, 700–930°F). This is essential for rapid crust formation so that steam trapped in the interior causes a "pocket" to form. In contrast, pumpernickel bread is often baked at a temperature as low as 100°C (212°F) for up to 24 hr. Optimal baking conditions are product-specific, and the best way to determine them is often through trial and error on a small scale.

COOLING

Proper cooling of loaves once out of the oven is essential to ensuring that condensation does not form on the inside of a package. Additionally, if the product is to be sliced, this cannot be accomplished easily until the product has cooled and the structure has set, at least to some degree. Cooling to about 38°C (100°F) is usually sufficient to meet these requirements. Commercially, cooling is usually achieved through the use of chambers or tunnels equipped with exhaust fans to draw in fresh air and exhaust the heat and moisture radiating from the cooling loaves. Equipment using vacuum processes, which enable faster cooling, and controlled atmosphere processes, which enable better control of cooling rates, is also available.

Product Issues

APPEARANCE

One of the most obvious traits of a baked product is the golden brown color of the crust. This color results from polymerization reactions known as Maillard browning and *caramelization*. Maillard browning occurs when amine groups on amino acids combine with the carbonyl groups of reducing sugar molecules. It is temperature- and pH-dependent, with higher pH increasing the reaction rate. The reaction continues, and colored pigments known as melanoidins eventually form. Caramelization involves only the sugars in the system, and, although it is fostered by conditions of higher temperature and lower moisture than Maillard browning, it likely contributes to the appearance as well.

Consequently, the important primary factors to consider when addressing a problem with browning are temperature, moisture, pH, reducing sugar content, and protein or amino acid content. It is important to determine which of these factors has changed and thus created the problem and then to return it to its previous condition. Lowering the bake time or temperature of a product always decreases browning, but it may also compromise some other quality characteristic (e.g., texture).

Another appearance factor that is very apparent is the dullness or shininess of the surface. A starch gel on the surface generally yields a smoother, shinier surface. Hence, factors relating to gelatinization, such as moisture content and temperature, are important. For products in which crustiness or shininess is important, the amount of water on the surface during baking is critical. Processing factors such as when and how much steam is injected into the oven should be carefully controlled.

Large blisters on the surface of a product or large voids in the interior crumb are the result of air trapped under the surface of the dough. This usually occurs when dough is sheeted, folded, and sheeted again. Processing to eliminate blisters includes a means of ensuring that air is not trapped initially or provision of a means for it to escape. Ensuring that dough pads are rolled tightly is important. In some cases, it may be necessary to mechanically penetrate (i.e., dock) the dough to allow the air to escape.

TEXTURE

A dough is a complex system, and many problems associated with the poor textural quality of a final product can result from a deficiency in one or more of the following dough characteristics: gas generation, gas retention, and setting of the structure in the expanded state. If problems are approached by focusing on these factors, complex problems can usually be solved. For example, *specific volume* can

Caramelization—The reaction occurring when sugars dehydrate and polymerize at high temperatures, producing brown pigments and flavor compounds that contribute to the flavor profile of many wheat-based products.

Specific volume—Volume in cubic centimeters divided by weight in grams.

be depressed if the leavening action is too low, if the dough is not optimally mixed (and therefore has poor gas-holding abilities), or if the gluten network is too elastic, making the side walls collapse on cooling. It is important to approach problems systematically, eliminating factors that do not affect the issue and rapidly identifying those that do.

FLAVOR

A bread product that is very bland and has a raw-flour flavor has probably been formulated either without salt or with insufficient salt. Other flavor issues are usually related to fermentation or the impact of other nonflour ingredients. One problem that often occurs in yeast-leavened products is the production of bread with a sour flavor. For sourdough breads, this is desirable. It is caused by the action of bacteria that produce lactic acid, which lowers the pH, creates sour flavors, and can affect leavening. Flour can contain significant levels of lactic acid bacteria, but the usual source of contamination for normal bread systems is the yeast. Bacteria grow well under the same conditions as those employed in yeast production; consequently, the yeast supplied for use as an ingredient is usually contaminated with lactic acid bacteria.

Yeast plays a role in flavor development in other ways as well. The development of fruity flavors can be the result of yeast. This is usually strain-specific and can be eliminated by using another yeast supply. If the yeast is allowed to act on the dough too long, the yeast cells can lyse (rupture), and the contents of the cell are liberated into the dough. Yeast proteases can hydrolyze peptide bonds, and the peptides formed in this manner can enter into reactions that lead to a "cheesy" flavor. The yeast also produces ethanol, which is a carrier for many flavor compounds and, if present in excess, can accentuate a flavor that is already there. It may also be perceived simply as an "alcohol" flavor.

Other ingredients that can affect flavor are chemical leavening agents. When an acidulant reacts with sodium bicarbonate, a residual salt is a by-product, and in some cases, this salt can be tasted. Other ingredients that affect bread flavor include sugars and malt. The Maillard and caramelization reactions also produce flavor precursors. In addition to color, the polymers produced in these reactions often carry bitter notes.

SHELF LIFE ISSUES

The most challenging issue associated with the keeping quality of bread products is staling. In the broadest sense of the term, "staling" affects both texture and flavor and both crumb and crust. A complete solution to bread staling has not been identified. With respect to firming of the crumb, however, some methods of solving staling, at least partially, do apply.

Firming of the crumb has been the subject of scientific study for over a century. Early work indicated that the degree of starch crystallinity correlated well with firmness. Later, Thomas Schoch postulated that the structure of fresh bread consisted of a firm amylose gel surrounding pliable, gelatinized, amylopectin-rich starch granules. Firming occurred as the end chains of the amylopectin molecules associated (Fig. 6-1). The granules became firm as a result, and the state of the starch in stale bread was firm granules surrounded by a firm amylose gel. Later, Hoseney and co-workers (3) hypothesized that gluten must also play a role in the firming of the crumb, since it is a continuous phase in bread. Starch is a discontinuous phase and, as such, should not affect the texture as much. In this theory, amylose exuded from the starch gels associates with the gluten. Increased associations over time bind the entire crumb structure together better and yield a firmer crumb (Fig. 6-2). These two concepts are not mutually exclusive. It is likely that the amylopectin in the granules in the latter model does associate, become more crystalline, and affect texture as well. Based on this understanding, it is clear that factors that interfere with starch-gluten interactions decrease staling. The use of α-amylase is effective because it severs the amylose chains connecting granules and gluten fibrils. Several patents (e.g., U.S. 5,059,430 and 5,209,938) concern the use of α-amylase to decrease staling. Monoglycerides and dextrins also interfere with these associations and can reduce the rate of firming.

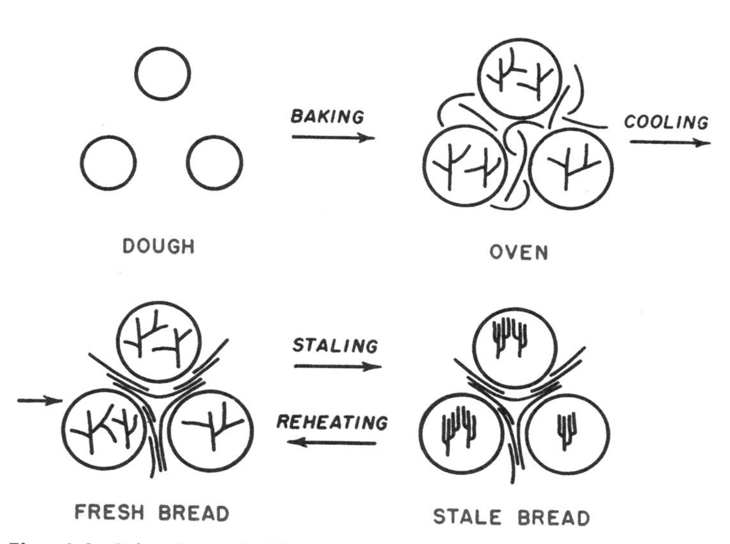

Fig. 6-1. Schoch model for bread crumb staling. Circles = amylopectin-rich starch granules after baking, branched lines = amylopectin molecules, unbranched lines = amylose gel. (Reprinted, with permission, from [2])

Another phenomenon often associated with staling is the formation of a tough, leathery crust with time. This results from moisture migration from the crumb to the crust. If a bread crust becomes leathery, it can be restored to a hard, crispy crust by reheating the bread briefly to dehydrate the crust again. Of course, this process can be repeated only a limited number of times before the entire loaf becomes dehydrated and overbaked.

Mold growth is another common mode of failure for stored bread products. Bread just out of the oven

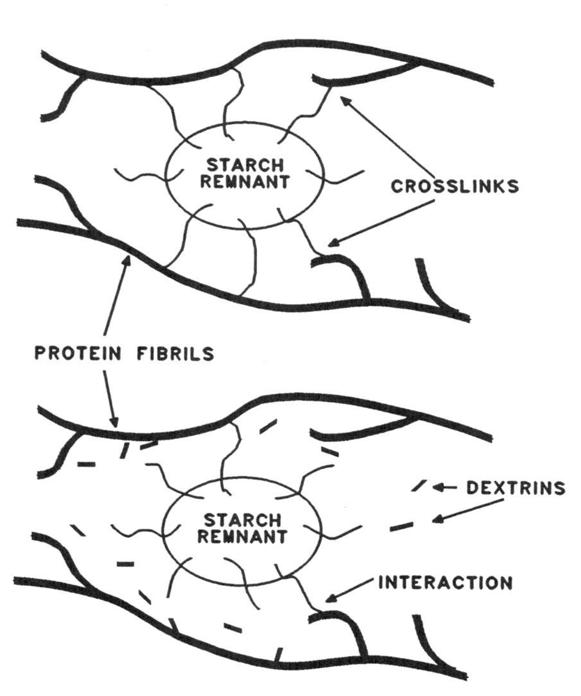

Fig. 6-2. Hoseney model for bread crumb staling. (Reprinted from [4], as adapted from [3])

is nearly sterile, but it can be contaminated with mold spores during cooling. Packaging does help, but, because mold spores are airborne, it is impossible to prevent molds though the use of packaging, which will be opened and closed during the consumption of the product. Hence, most companies use mold inhibitors such as calcium propionate in their formulations.

Processing Issues

Many modern dough-processing lines run at high speed, producing product at rates of up to 600 pieces per minute or higher. This usually allows much economic efficiency and, in many instances, a high degree of product consistency. However, these processes usually do not provide time for dough to relax, and this can lead to significant processing problems. Gluten is a polymer with very high molecular weight. When dough is mixed, gluten fibrils become entangled, and the mass becomes elastic as a result. Given time, these entanglements relax, and the dough becomes less elastic. If the dough cannot relax, however, and further work is applied, the gluten structure remains elastic. This can lead to tearing and snap-back problems.

For example, malformed products in processes in which the product is cut into a desired shape are usually caused by variation in the viscoelastic properties of the dough. If a dough is too elastic and not given enough time to relax, it snaps back when cut or divided. Furthermore, if it has been sheeted in one direction only, it snaps back more in the direction of the sheeting because the gluten polymers have been so aligned. This causes the product to be misshapen (e.g., products cut round become "football" shaped). Tearing usually occurs in forming processes if the work input after mixing exceeds the ability of the dough to absorb it. Additional work can be applied to a relaxed dough, and the dough does not usually tear. Work put into a fully developed elastic dough, such as forcing it through the gap of sheeting rollers, can cause the dough to tear. Consequently, important factors to consider when addressing shape or tearing problems are the basic viscoelastic properties of the dough, the effect of time on these viscoelastic properties, the amount of work applied to the dough after mixing, and the directionality of the dough.

Another problem related to viscoelastic properties of dough is layer inconsistencies in laminated products such as croissants. In the processes used to create these products, it is very important that the consistencies of the dough and the shortening or butter be similar to ensure that the layers are continuous and integral. If the butter or shortening is too hard, it "breaks through" the dough layers. Alternatively, if the dough is too stiff or elastic, the shortening layer will not be uniform. In addition to formulation, temperature control during processing is important because it affects the consistency of both the dough and the roll-in fat components.

Fundamental Mechanism of Breadmaking

A comprehensive, fundamental understanding of what occurs during dough development and baking can be very useful when attempting to identify solutions to product problems. The process begins in the mixer when the particles of flour hydrate slowly from the outside toward the center. Because of the shearing forces involved, the hydrated layers are sloughed off the outside of the particles, exposing those below, which, in turn, become hydrated and are later sloughed off also. This process continues until all of the flour particles are hydrated. Facilitated by mixing, the gluten fibrils from these particles interact, entangle, and eventually form a continuous three-dimensional network. This is the point of optimum gas retention and the peak of a curve from a recording dough mixer or wattmeter. Starch, the other minor components of the flour, and the other ingredients are hydrated and incorporated within the dough mass as well. Soluble compounds such as sugar and salt go into solution. Air cells also form in the dough during mixing as a result of the folding and stretching action. At this point, a dough is optimally mixed, but if there is further work input, bonds in the gluten break; the matrix is less complete; and the dough is less elastic and less able to retain gas.

The enzymes in the system are also hydrated and become active. α-Amylase begins to act on the damaged starch in the dough mass. The resultant sugars, the sugars native to the flour, and the added sugars dissolve and diffuse to the hydrated yeast, which begins to produce carbon dioxide, ethanol, and compounds that ultimately contribute to the bread flavor. At first, the carbon dioxide goes into solution, and ionic carbonate and bicarbonate concentrations begin to rise. Eventually, as the water becomes saturated with carbon dioxide and related carbonate species, these gases diffuse to the boundaries of the air cells and then into the cells. The rate of diffusion of carbon dioxide and related species though the dough determines the gas-holding ability of the dough. As the reaction continues, the dough expands (i.e., proofs). Its physical properties change, making it capable of being stretched into very thin sheets (Fig. 6-3). The cells become larger and richer in carbon dioxide, and the cell walls become thinner as the dough expands. If the dough is punched, gas is expelled and the cells are divided, producing more cells. Then, as the yeast continues to produce gas, the cells grow, and the dough expands again.

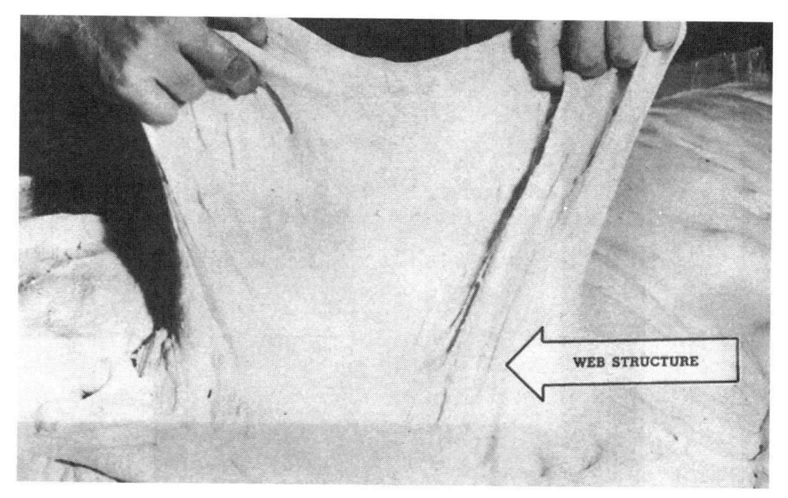

Fig. 6-3. Characteristics of dough after bulk fermentation. (Reprinted, with permission, from [5])

Web Sites

American Institute of Baking—www.aibonline.org

The American Institute of Baking is a not-for-profit corporation, founded by the North American wholesale and retail baking industries in 1919 as a technology transfer center for bakers and food processors. The original mission of the organization was to "put science to work for the baker," and that basic theme is still central to all of the programs, products, and services provided by AIB to baking and general food production industries worldwide.

National Baking Center—www.nationalbakingcenter.com

The National Baking Center's goal is to support a renaissance in traditional baking. The Center's mission is to raise the level of professionalism and quality in American traditional baking through education and service to the industry.

The Bread Baker's Guild of America—www.bbga.org

The organization's mission is to provide education in the field of artisan baking and the production of high-quality bread products.

The Retailer's Bakery Association—www.rbanet.com

RBA is an organization of businesses producing and selling bakery foods at retail, along with their suppliers. RBA is a resource center for the development of knowledge and skills for the retail bakery industry. It creates economic leverage for independent retail bakeries.

When the dough has achieved the desired volume, it is placed in an oven and baked. Because heating occurs from the outside to the center, the starch on the surface gelatinizes first. As the dough warms, the air cells expand because the volume that a gas occupies is directly related to its temperature. The ethanol-water *azeotrope* vaporizes and contributes to the leavening gases. Until inactivated by heat, yeast also produces gas at a higher rate. Under the influence of these leavening gases, the dough expands further, yielding the phenomenon known as oven-spring. As heat penetrates, the starch gelatinizes and becomes very hygroscopic; the gluten network dehydrates; and a major change occurs in the water balance of the system. Moisture is driven to the surface and evaporates into the oven atmosphere, keeping the surface relatively cool. In the dough, amylose is exuded from the starch granules, but the granules remain largely intact because there is not enough water available to fully paste the starch.

As the air cells become larger, the gluten in the cell walls is stretched, *strain hardens*, and becomes brittle. Eventually, the cell walls rupture, and the "air-discontinuous" dough becomes "air-continuous" bread. The starch has fully gelatinized, and there is less free water to diffuse to (and evaporate from) the surface. Without the evaporative cooling effect, the surface temperature rises, and the sugars and amine groups in the crust react to form the precursors to melanoidins, which are brown pigments. The reaction continues and, as the polymerization

Azeotrope—A miscible mixture of solvents that cannot be separated by boiling.

Strain hardening—Becoming more resistant to extension and harder as a result of being stretched.

reactions of Maillard browning progress, the color of the crust changes from yellow to golden brown. Also contributing to browning is caramelization, a polymerization reaction involving only the sugars in the system. A hard crust forms because the surface dehydrates more than the interior. Upon cooling, the viscosity of the starch gel in the crumb rises and the structure sets. Because the air cells are all interconnected, there is no pressure differential between air in individual cells and the external atmosphere to drive a collapse of the bread structure.

Troubleshooting

This section lists some common problems, causes, and suggestions for changes to consider in formulation and processing. However, any flour-based product is complex, and a simple solution is not always possible. Consequently, a solution may involve one or more of these factors or others specific to the system in question. This guide may serve as a starting point for the solution to a problem, but a solution based on fundamental knowledge and/or experience with the product generally leads to a better, longer-lasting resolution to the problem.

APPEARANCE		
Symptom	**Possible Causes**	**Changes to Consider**
Too brown	High sugar content	Reduce formula sugar, especially reducing sugars. Check flour α-amylase. Check starch damage.
	High bake temperature	Lower oven temperature.
	Long bake time	Reduce bake time.
	Low moisture	Increase formula water.
	High pH	Check dough pH.
Blistered surface	Air pockets under dough surface	Adjust molder. Dock dough. Reduce dusting flour.
Voids in crumb	Air pockets in dough	Check process during rolling, molding, or folding operations. Reduce dusting flour.
Dull surface	Underdeveloped starch gel on surface	Adjust steam during baking.

TEXTURE		
Symptom	**Possible Causes**	**Changes to Consider**
Low volume or dense crumb		Poor gas-holding properties. Adjust mix time. Check post-mixing work input. Adjust flour and/or water content. Change flour type. Eliminate reducing agents. Add oxidant. Add dough strengthener. Add vital wheat gluten.
	Inadequate leavening	Increase yeast content. Ensure that adequate sugar is present. Increase chemical leavening content. Ensure correct balance of acidulant and bicarbonate.
	Structure does not set in expanded form	Remove gluten. Use weaker flour. Add malt. Add reducing agents. Check oven temperature.

FLAVOR		
Symptom	**Possible Causes**	**Changes to Consider**
Bland	No salt or low salt	Check formula salt level.
Sour	Lactic acid bacteria	Check yeast for contamination.
Cheesy	Protease activity	Reduce fermentation time. Eliminate added proteases.
Fruity	Yeast type	Change yeast type. Reduce fermentation time.
Bitter	Browning reactions	See appearance issues above.
Chalky, salty, or metallic	Chemical leavening by-products	Change acidulant.

SHELF LIFE		
Symptom	**Possible Causes**	**Changes to Consider**
Crumb firming	Interactions between starch and gluten	Add α-amylase. Add dextrins. Add monoglycerides.
Leathery crust	Moisture migration from crumb	Reheat.
Mold	Contamination with mold spores	Add preservative.

PROCESSING		
Symptom	**Possible Causes**	**Changes to Consider**
Malformed product	Viscoelastic properties	Check mixing time. Allow more rest time. Reduce directionality in molding or sheeting operation. Adjust flour strength. Add dough strengthener or reducing agent.
Tearing	Viscoelastic properties	Check mixing time. Allow more rest time. Adjust flour strength. Add dough strengthener or reducing agent.

References

1. Milling and Baking News/Baking and Snack, 1999-2000 Reference Source. Sosland Publishing Co., Kansas City, MO.
2. Schoch, T. J. 1965. Starch in bakery products. Baker's Dig. 39(2):48.
3. Martin, M. L., Zeleznak, K. J., and Hoseney R. C. 1991. A mechanism of bread firming. I. Role of starch swelling. Cereal Chem. 68:498-503.
4. Hoseney, R. C. 1994. *Principles of Cereal Science and Technology*, 2nd ed. American Association of Cereal Chemists, St. Paul, MN. Chapter 12.
5. Swortfiguer, M. J. 1950. White bread sponge and dough development. Proc. Am. Soc. Bakery Eng. p. 94.

Supplemental Reading

1. Pyler, E. J. 1988. *Baking Science and Technology*, 3rd ed., Vols. 1 and 2. Sosland Publishing Co., Merriam, KS.

CHAPTER 7

Products from Soft Wheat Flour: Problems, Causes, and Resolutions

In This Chapter:

Ingredients
- Flour
- Chemical Leavening
- Shortening
- Sucrose
- Egg Whites

Formulation
- Cakes and Related Products
- Doughnuts
- Crackers
- Cookies
- Biscuits
- Pie Crusts

Processing
- Cakes and Related Products
- Doughnuts
- Crackers
- Cookies
- Biscuits
- Pie Crusts

Product and Processing Problems
- Cakes
- Doughnuts
- Crackers
- Cookies
- Biscuits
- Pie Crusts

Troubleshooting

The diverse products made with soft wheat flour include the general categories of cakes, doughnuts, crackers, cookies, biscuits, and pie crusts. Unlike hard wheat products, which are typically breads or bread-related products, each soft wheat product is a unique system. For example, the water relationships involved in processing vary dramatically through the range of soft wheat products—from batters, which are used to produce cakes, waffles, muffins and related products, to very stiff doughs such as those for cookies and pie crusts. One commonality of soft wheat products, however, is that most of them are chemically leavened, unlike hard wheat products, which are usually yeast leavened.

Ingredients

FLOUR

The flour used in cake products baked in the United States is invariably chlorinated soft wheat flour. The chlorine affects all components of the flour, but the oxidizing effect of chlorine on the starch affects its performance most significantly. When starch is chlorinated, it is able to swell to a greater extent and hence raises the viscosity of a batter more than does an unchlorinated flour. A by-product of flour chlorination is hydrochloric acid, which depresses the pH of the flour. The pH of most cake flour is 4.7–4.9. In fact, pH is used as a common analytical measure of the extent of flour chlorination.

The flour used in cookies is usually not highly chlorinated. The water-binding properties of a cookie flour are very important to controlling quality in terms of cookie appearance and size. For this reason, starch damage should be kept to a minimum, and flour-water relationships should be carefully controlled. Crackers, biscuits, and pie crusts are usually produced with an unchlorinated soft wheat flour. In many soft wheat products, bran specks on the surface of the product are undesirable. In these cases, a low-extraction flour or a flour made from soft white wheat should be used.

Encapsulation—A technology that enables small particles or droplets of materials to be coated, usually with a fat.

CHEMICAL LEAVENING

The fact that most soft wheat products are chemically leavened is a distinguishing difference from hard wheat products in that the leavening affects both flavor and texture. Several different chemical leavening systems are used in soft wheat products. They vary in how fast they react to produce carbon dioxide (i.e., at what point or points in the processing) and what residual salts are left after the reaction. The residual salts affect the flavor of the final product. Most, but not all, chemical leavening systems contain sodium bicarbonate as the primary source of the leavening gas carbon dioxide. Potassium bicarbonate can be used for this purpose in formulas in which high sodium content must be avoided. An acidulant is usually used in combination with the bicarbonate source; this lowers the pH of the system and causes the evolution of gaseous carbon dioxide. An exception to this rule is ammonium bicarbonate, which evolves ammonia, water, and carbon dioxide when heated without an acidulant. Ammonium bicarbonate leaves no residual salt to affect flavor, but products using this system must be baked to near dryness (i.e., 2–3% moisture) to ensure that all the ammonia gas has been driven out of the product. If not, residual ammonia causes off-flavors in the product.

Because chemical leavening systems usually contain acids and bases, the reaction rate, and hence the rate of gas evolution, can be controlled by influencing the reactivity of either component. *Encapsulation* is one means of reducing the reactivity of the bicarbonate, and it can be used to delay the reaction of a fast-acting acidulant as well. More commonly, however, the rate of reaction is controlled by selecting the appropriate acidulant to use in combination with free sodium bicarbonate (Table 7-1).

TABLE 7-1. Reaction Rates and Neutralization Values of Common Acidulants[a]

Acid	Formula	Neutralization Value	Relative Reaction Rates[b]
Cream of tartar (monopotassium tartrate)	$KHC_4H_4O_6$	45	1
Monocalcium phosphate monohydrate	$CaH_4(PO_4)\cdot H_2O$	80	1
Anhydrous monocalcium phosphate	$CaH(PO_4)$	83.5	2
Sodium acid pyrophosphate	$Na_2H_2P_2O_7$	72	3
Sodium aluminum phosphate	$NaH_{14}Al_3(PO_4)_8\cdot 4H_2O$	100	4
Sodium aluminum sulfate	$Al_2(SO_4)_3\cdot Na_2SO_4$	100	4
Dicalcium phosphate dihydrate[c]	$CaHPO_4\cdot 2H_2O$	33	5[c]
Glucono-δ-lactone	$C_6H_{10}O_6$	50	...[d]

[a] From (1).

[b] Relative rate: 1 = reactive at room temperature, 5 = requiring oven temperature for reaction.

[c] Generally reacts too slowly to be a leavening acid; used to adjust final pH.

[d] Reaction rate depends on many factors in addition to temperature.

An acidulant should be used at a level such that it exactly neutralizes the sodium bicarbonate to produce the optimal amount of carbon dioxide without affecting the pH of the product. The neutralization value of an acidulant can be used to determine the correct balance of sodium bicarbonate and acidulant. It is found with this equation:

$$\text{Neutralization value} = \left(\frac{\text{g of sodium bicarbonate} \times 100}{100 \text{ g of acidulant}}\right)$$

For example, if the neutralization value of an acidulant is 80, then 80 g of sodium bicarbonate is required to neutralize 100 g of the acidulant. See Box 7-1.

Box 7-1. A Note on Acidulants

In some products, acid or basic pH levels are desirable and the acidulant and sodium bicarbonate are not balanced. Examples include chocolate-containing products and old-fashioned cake doughnuts. Additionally, ingredients (e.g., buttermilk) in some formulations affect pH and thus necessitate variation from an exact balance of bicarbonate and acidulant.

Solubility is related to reactivity since the acidulant must be in solution to react. The most soluble and most reactive acidulant is monocalcium phosphate (MCP). When this acidulant is used in conjunction with free sodium bicarbonate, most of the leavening gas is evolved in the mixer. Cream of tartar (i.e., the potassium salt of tartaric acid) is similar in reactivity to MCP but is not commonly used because of its expense. Glucono-δ-lactone (GDL) is another fairly fast-acting acidulant. Unlike the salts of the phosphate-based acidulants, GDL's residual salt is a derivative of glucose and has less effect on the flavor profile of the finished product. If it is used with free sodium bicarbonate, some proofing occurs on-line, even during a high-speed process. Slower-acting acidulants include sodium acid pyrophosphate (SAPP) and sodium aluminum phosphate (SALP). These are available in several grades and reactivities. SAPP has a strong and distinctive metallic aftertaste (from residual pyrophosphate) that can be undesirable, whereas SALP has a much cleaner flavor. Neither of these acidulants can leaven a product significantly until some heat is applied. This is the case even in batter systems. The slowest-acting common acidulant is dicalcium phosphate (DCP). DCP requires high heat to react and is often used in chemical leavening systems containing more than one acidulant to give a secondary gas-evolving "boost" once the product is in the oven. The bicarbonate can often be combined in a dry form with one or more acidulants to form *baking powder*. A single-acting

Baking powder—A preparation containing sodium bicarbonate and one or more acidulants.

baking powder has one acidulant. A double-acting baking powder has one acidulant that reacts in the dough or batter and another that reacts when heated.

SHORTENING

Shortening used in batter-based products usually contains monoglycerides and diglycerides as *emulsifiers*, which aid in the dispersion of the shortening throughout the batter. In cookies, unemulsified shortening is used, and shortening level can influence the spread or surface impressions of the final product by reducing viscosity during baking. In the formulation of all soft wheat products, shortening is added primarily to provide a soft, tender bite. If it is low or absent, the product more closely resembles a bread product.

SUCROSE

One obvious reason for a high level of sucrose and other sugars in soft wheat products is that they function as flavoring agents. Sucrose plays other roles in many systems, however, one of which is as a *plasticizer*. In this role, it actually provides more effective liquid to a system. When 1 g of sucrose dissolves in 1 mm of water, 1.6 mm of solution results. This can have the effect of liquefying a solid system. For example, if 1 g of sucrose is added to a mixture of 1 g of water and 1 g of starch, the mixture is converted from a powder to a suspension (Fig. 7-1). Another role of sucrose is to delay gelatinization. In some

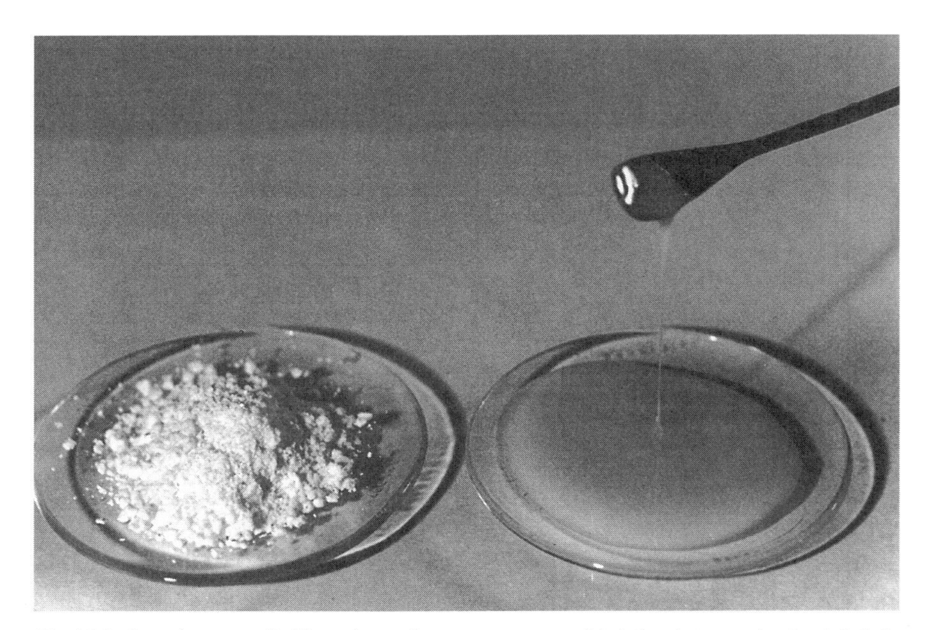

Fig. 7-1. Starch-water (left) and starch-sucrose-water (right) mixtures, both at 1:1:1, illustrating the transformation of a powder to a suspension by the addition of a dissolved solid. (Reprinted from [2])

Shortening—Solid fat derived from plants.

Emulsifier—An additive that aids the dispersion of fat into a formula.

Plasticizer—An ingredient that renders a mixture more flowable or pliable.

soft wheat products (e.g., cakes), starch gelatinization plays a major role in setting the structure. Sucrose has the effect of delaying gelatinization so that the structure sets after the leavening action has taken place.

Foam—A gaseous noncontinuous phase held in a continuous phase.

EGG WHITES

The proteins of egg white have the ability to form films, hold air incorporated during mixing, and thermoset during baking. In angel food cakes, the *foam* produced by whipping the sugar and egg white is the primary structure of the batter, and it must be preserved or the cake will fail. For this reason, it is important to gently incorporate the flour and avoid the use of shortening to ensure that the foam is not destabilized.

Formulation

CAKES AND RELATED PRODUCTS

Cakes, muffins, quick breads, cake doughnuts, pancakes, and waffles are all made from batters. Batters contain more water than doughs do, but the distinction between a dough and batter is not always clear. Although both doughs and batters flow, a batter generally can be poured from one container to another in a reasonable amount of time. Each of the products above requires specific formulation and processing. However, the layer cake is perhaps the most representative system in this category.

A layer cake formula contains a high amount of sugar (Table 7-2). The term "high ratio cake" indicates that the formula has more sugar than flour. Other ingredients critical to a layer cake formula are baking powder, emulsified shortening, a heat-coaguable protein (e.g., egg

TABLE 7-2. Typical Formulas for Three Types of Cake[a]

	Percent (Flour Basis) for		
Ingredients	Rich White Layer Cake	Angel Food Cake	Commercial Pound Cake
Flour	100	100	100
Sugar	140	500	100
Shortening	55	...	50
Eggs			
Whites (fresh)	76	500	...
Whole (fresh)	...	...	50
Milk (fresh)	95	...	50
Baking powder	1.3	...	...
Cream of tartar	...	20	...
Salt	0.7	...	...

[a] From (1).

Surfactants—Surface-active agents that affect how two materials with different properties interact at the surface between them.

whites), and water or milk. *Surfactants* (e.g., propylene glycol monostearate) are often added to aid in the incorporation of air into the batter. Another popular cake, the pound cake, is generally formulated with about equal amounts of flour and sucrose, as well as shortening and whole eggs but no leavening. In this product, the air cells incorporated during mixing provide the only leavening. Angel food cakes generally contain up to five times as much egg whites and sucrose as flour, and only the acidulant cream of tartar is added to control the pH.

Muffins and quick breads are made from variations of cake formulas that usually have much lower sugar levels (about 5–15% of the total formula). These products vary widely in their minor ingredients; many include fruit bits, bran, or other particulates as flavoring agents. Waffle and pancake batters are made from simple formulas that are usually composed only of flour, water, baking powder, and small percentages of sugar, oil or shortening, and flavoring agents.

DOUGHNUTS

All doughnuts are fried, not baked, and therefore frying oil is an ingredient, one not common to many other soft wheat products. A cake doughnut formula is very similar to a layer cake formula, typically with 12 ingredients or more. It contains less water than a cake batter, however, and the batter is more viscous. A yeast-raised doughnut is made from a sweet dough, which is allowed to ferment. Yeast-raised doughnut formulas often include yeast foods and dough conditioners not present in cake doughnut formulas. Both straight-dough and sponge-dough processing are used for the fermentation.

Because of the popularity of doughnuts and the need to produce them on a daily basis, many bakeries purchase preassembled mixes that contain all the base ingredients. Then, if additional ingredients are desired, they are added to the mix. This avoids the time-consuming process of scaling all the individual ingredients in the formula. It also increases the uniformity of the mix and, consequently, the quality and uniformity of the product.

CRACKERS

There are several types of crackers, but all are made from doughs. Water relationships are very important in cracker processing, and soft wheat flour with a low and constant absorption is preferred. Crackers are made from low-moisture doughs that generally contain low amounts of sugar and high levels of fat. Some crackers, such as saltine crackers, contain yeast in the formula in addition to sodium bicarbonate to reduce the elasticity of the dough and provide the characteristic flavor. Snack crackers usually contain no yeast, and they have higher amounts of shortening and more ingredients affecting flavor than saltine crackers.

COOKIES

These soft wheat products are known as "biscuits" in most countries other than the United States. Soft wheat flour is preferred for the production of cookies because it binds less water than hard wheat flour. Water relationships in a cookie dough have a major effect on cookie quality (e.g., cookie spread, texture). If hard wheat flour is used in a cookie formula, the result is usually a tough or very hard cookie that spreads very little during baking. Cookies are made from a very wide range of formulas. Generally, they contain high amounts of flour, shortening, and sucrose and little water. All cookies meeting this general formula are made from doughs. Wafer-type cookies and related products such as ice cream cones are made from batters and hence are the exception. Wafer formulas contain no shortening or sugar and excess water. Specks apparent in wafer products are considered very detrimental. Consequently, low-extraction flour or flour made from soft white wheat is preferred.

BISCUITS

A "biscuit" in the United States is a small, chemically leavened bread. Soft wheat flour is preferred to give the biscuit a short texture, but when high-speed processing is required, hard wheat flour is often substituted. Most biscuits are made from a dough similar in consistency to a bread dough. Others (i.e., drop biscuits) are made from a significantly wetter dough, one closer in consistency to a batter.

PIE CRUSTS

The primary ingredient in a pie crust besides flour is shortening or *lard* (about one-third of the total formula). Salt is also an important ingredient with respect to the flavor of the final crust. The dough is very low in moisture, and much of its consistency is dependent on the consistency of the shortening or lard. Many manufacturers of pie crusts prefer lard because it imparts a characteristic flavor and texture to the finished pie crust. It is important to minimize work input during mixing. The optimal dough has "islands" of lard so that, when it is sheeted, sheets of lard separate cereal layers, thus yielding flakiness.

Processing

CAKES AND RELATED PRODUCTS

The simplest continuous layer cake process, among the many variations that exist, includes a slurry mixer, holding tank, continuous mixer, depositing system, and oven. Ingredients are combined and hydrated in the slurry mixer. This mixing is done at low speed and does not aerate the system. The batter is then pumped to the holding tank, which is a reservoir for the continuous mixer. Aeration is effected

Lard—Solid fat derived from pigs.

in the continuous mixer, where the batter is whipped at high speed and high shear (up to 300 rpm) under significant air pressure (135 psi). The mixing action causes the batter temperature to rise if it is not controlled. Batters are usually kept cool during this process by a jacket of cold water or by altering the speed of the mixing blades. The batter is then pumped at low pressure and deposited into cake pans. The pans are conveyed to an oven, where the batter is transformed to a cake.

Other products such as muffins and quick breads do not require the aeration a layer cake requires; hence, the high-speed mixing is absent. A low-speed mix is employed to avoid separation of the gluten from the rest of the ingredients. As with cakes, these batters are deposited in baking vessels and transported through an oven.

Mass-produced pancakes are made by a process in which batter is deposited on heated surfaces. The pancakes are then mechanically flipped. They are usually frozen for distribution. Waffles, a similar product, are baked in ovens with rotating pairs of plates. The bottom plate is filled with the batter; the top plate is closed; and the batter is enclosed between them as it is transported through the oven.

All of these products are also made by the consumer at home from retail prepared mixes. The preparation instructions included with the mixes simulate the processes above. For example, to achieve the batter aeration required for a layer cake, the consumer is instructed to beat the batter with a hand mixer for a specified length of time. Even so, these products are more dependent on leavening and emulsification than on the direct air incorporation used in commercial processes. These products are declining in popularity as the amount of mixes sold to the foodservice industry rises.

DOUGHNUTS

Cake doughnuts are made from a stiff batter. The ingredients are mixed gently, and the batter is allowed to rest 15 min so that it is fully hydrated. The batter is then transferred to a hopper, from which it is deposited in the appropriate form directly into a fryer. The doughnuts are then cooled and coated with powdered sugar, icing, or other flavoring agents.

Doughs for yeast-raised doughnuts are mixed with the sponge-dough process. The doughnuts can be formed by sheeting and cutting the dough into an appropriate shape or by extruding it with a low-pressure extruder. As with cake doughnuts, yeast-raised doughnuts are allowed to cool before coatings are applied.

CRACKERS

Saltine crackers are made from a sponge-dough mixing operation (Fig. 7-2). The fermentation of the stiff, low-moisture sponge is unique in that it is generally very long (i.e., about 16 hr). Both yeast and bacterial fermentations are important to the success of the product. The final dough is further fermented for about 6 hr and then allowed to relax. To achieve the desired flakiness, the dough is sheeted, folded, and turned 90° to achieve a final dough sheet that has about eight very thin layers and has been sheeted in both directions. The dough sheet is then cut, docked, and transported through a high-temperature (250°C, ~480°F), direct-fired oven on a mesh band for a short time (about 2–3 min). The crackers are then cooled slowly to avoid cracking.

Snack cracker processing is less complicated. A straight-dough mixing process is employed. The mixed dough is then sheeted with little or no fermentation, laminated (if flakiness is desired), cut, docked, and baked. Often, these crackers are cut into circles or other shapes that create a large amount of scrap dough that must be recycled to the mixer.

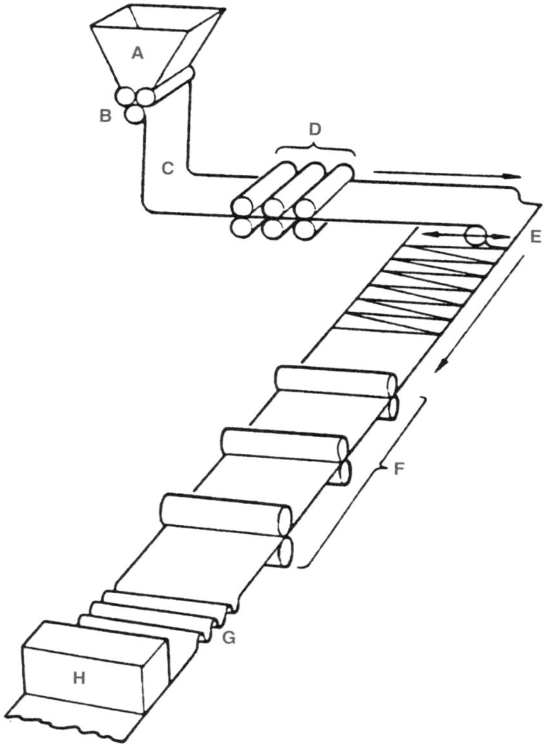

Fig. 7-2. Steps in the processing of saltine cracker dough. A = dough hopper, B = forming roll, C = dough web, D = reduction rolls, E = lapper, F = final reduction rolls, G = relaxing curl, H = cutting and docking. (Reprinted from [1])

COOKIES

The four general processes used to make cookies include rotary molding, wire cutting, machine cutting, and wafer baking. Three of these are shown in Figure 7-3. Most cookies are baked in long tunnel ovens on solid bands, with the exception of wafers, which are molded and baked between plates.

In the rotary-molding process, a stiff, yet plastic dough is mixed in a single stage. It is then roughly sheeted and forced to engage a roll with molds of the cookie shape and design cut into its surface. The roller rotates, and, on the other side, the cookie dough is extracted from the mold directly to the oven band. It is then conveyed through the oven to bake. The consistency of the dough is very important to ensure that the cookie can be easily transferred into and out of the mold and that the design on

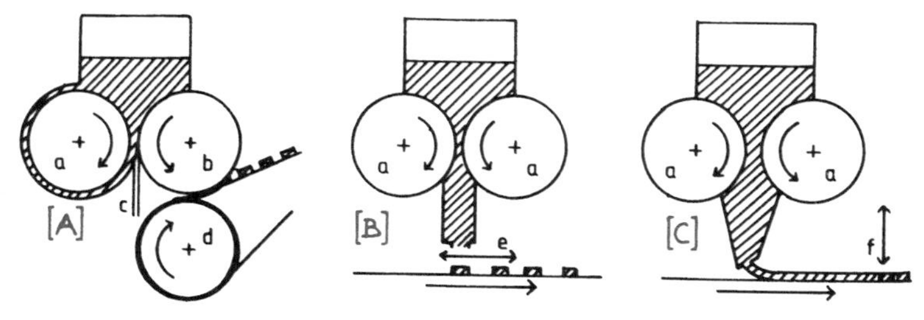

Fig. 7-3. Principles of operation of rotary (A), wire-cut (B), and bar or rout (C) presses. a = grooved roller, b = forming roller, c = oscillating knife, d = rubber-covered roller, e = wire, f = guillotine. (Reprinted from [3])

the cookie is not distorted. The dough is very dry (about 8% moisture) and not elastic.

Machine-cut cookies are similar to rotary-molded cookies, but the dough has more water and hence is not as stiff. The dough is developed during mixing and sheeting so that there is a continuous gluten matrix that inhibits rise and spread during baking. The cookie dough is then sheeted, stamped, cut, and baked.

Wire-cut cookies are extruded and immediately cut with a reciprocating wire. Consistency of the dough is important. It must be maintained so that it cuts cleanly but is also cohesive enough so that it does not crumble. The dough is generally softer than the rotary-molded cookie dough described above. It does rise and spread when baked; therefore, strict formula controls are required to ensure that the cookie dimensions are maintained within target.

Wafers are baked on equipment similar to the waffle oven described above (Fig. 7-4). Extracting the finished wafer from the plate is a primary problem encountered in this process, and nonstick surface coating of the plates is one solution. After the wafer is baked, it is cut into sheets. If the product is to be layered with filling, that is done at this point before it is cut to the final dimensions.

Fig. 7-4. Sugar wafer baking, showing the wafer sheet still in place on the baking plate. (Courtesy Franz Hass Machinery of America, Inc., Richmond, VA)

BISCUITS

Commercial biscuit processing is similar to bread processing. Biscuit mixers may be horizontal bar mixers or sigma-action mixers, depending upon the development and final texture desired. The dough is generally mixed in one stage, although a complex formula may be mixed in two stages, with salts and other hygroscopic ingredients added in the second stage. This ensures that the flour is hydrated and that the gluten develops into a continuous matrix. The dough is then sheeted and, in the simplest form, cut and baked. For flakier biscuits, shortening is layered in with alternating folding and sheeting operations.

PIE CRUSTS

Pie crust mixing is very important to achieving a flaky, high-quality product. Overmixing and high temperature are detrimental at this stage because, under these conditions, the fat tends to be dispersed evenly throughout and the finished crust becomes grainy. Working the dough after the mixer is also detrimental; hence, a simple sheeting operation produces optimal results. Commercial pie processes employ pie machines that automatically take sheets of dough, cut them into rectangles, and cross roll them to round them. Bottom crusts are then placed in pie pans, and the filling is deposited, followed by the depositing of the top crust. The crusts are then automatically crimped and trimmed. The unbaked pie is finally conveyed to a band oven.

Product and Processing Problems

CAKES

One of the most common problems associated with cakes and related products is final product volume. The final volume and consequent density of the crumb are dependent on the amount of gas incorporated into and generated in the batter, the ability of the batter to retain that gas, and the required setting of the structure in the expanded state. Focusing on these factors aids the solution of volume-related problems. Another related problem associated with cakes is their crumb grain. In most batter-based products, a fine grain is desirable. The grain of a cake is determined not by the total gas in the batter but by the distribution of the gas in the form of many small bubbles.

As is true for doughs, air bubbles are incorporated into batters during mixing and cannot be created later in the process. Hence, the amount of gas in the batter is largely dependent on the mixing process. If baking powder is present, however, some gas is generated in the batter after mixing. If air incorporation during mixing is the cause of a volume problem, then monitoring and controlling the density of the batter is critical. As more air is incorporated, the batter density goes down.

Retaining the gas in a batter is largely dependent on its viscosity. If the batter is too thin, bubbles coalesce, migrate to the surface, and are lost to the environment. Vibrations and/or sitting too long before baking can also foster this process and cause cakes to fail. Many ingredients and processing factors influence batter viscosity. Certainly, the water or milk level in the formula has an effect, but flour, sugar, emulsified shortening, and surfactants also significantly affect the viscosity of a batter. Strict control of the type and level of all ingredients in a cake formula is essential to maintaining product consistency.

Starch gelatinization and protein denaturation both contribute to the setting of a cake structure. As a batter is heated, viscosity drops dramatically, then levels off, only to rise again very sharply. The rise in viscosity (and hence the setting of the structure) is attributable primarily to starch gelatinization, but when egg whites are added, the viscosity is higher at all temperatures. Clearly, the denatured protein of the egg white is a structural component of the final cake as well. Another factor to be considered in controlling the point (i.e., temperature and time) of starch gelatinization is the balance of the basic ingredients flour, sugar, and water in the formula. Sugar, because it is present in such high quantities, does delay gelatinization.

A problem that occurs in a low-viscosity batter is the separation and settling of the starch. Starch has a higher density than many batter components and settles if the viscosity is not high enough to keep it suspended. When heat is applied during baking, this settled starch forms a gel on the bottom of the cake, which is perceived as an undesirable gum layer.

Fine grain will be achieved if many tiny air bubbles are incorporated into the batter during mixing and maintained through baking. Surfactants such as propylene glycol monostearate or the lecithin in whole eggs affect the surface properties of the air bubbles and hence how they interact. If used at optimal levels, these additives increase viscosity, foster the incorporation of air in the mixer, and help maintain the dispersion of air cells through baking. Therefore, if grain is a problem, evaluation of the surfactants in a formula is warranted.

A problem common to many commercial batter processes is the separation of gluten. Agglomerated gluten can clog nozzles in depositors and accumulate in the bends of pipes. Soft wheat flour, in general, has low protein content and gluten that does not associate as well as the gluten of hard wheat flour. Regardless, selection of a lower-protein flour or dilution of the flour with wheat starch may be advantageous as long as product quality parameters are not compromised. The best way to address these problems, however, is to minimize the work put into the batter by pumps, impellers, and other processing equipment.

Control of the pH of the cake is important to its quality. Cakes baked from low-pH batters can exhibit an undesirable acidic flavor, whereas a high-pH batter produces a cake with the flavor of a soda biscuit. Proper control of the ingredients influencing pH (i.e., flour chlorination and leavening agents) is critical. The pH of a batter can also influence viscosity, resulting in volume and texture effects. In chocolate cakes, pH can affect color as well. In general, higher-pH batters produce cakes with coarser grain and higher volume.

DOUGHNUTS

Overmixing or undermixing a doughnut batter or dough can result in product with poor texture and incorrect fat absorption. Overmixing deaerates the system and causes a dense, tough product. Undermixing does not adequately blend the ingredients, so the structure becomes weak and a very fragile doughnut results.

Fat absorption is also a factor important to doughnut quality. The different types of doughnuts have a fat-absorption range of about 20–30%. Moisture content plays a major role in the amount of fat absorbed by a flour-based product during frying. The lower the moisture content, the lower the amount of fat absorbed.

One indicator of how well the doughnut product has been formulated is the time the doughnut initially remains submerged in the fryer. Generally, the doughnut sinks and then rises after about 5 sec. This time is dependent on the formulation (e.g., leavening level) and condition of the batter or dough. Monitoring this simple parameter can help control the texture and fat absorption of the finished doughnut.

CRACKERS

Checking—The production of fissures in crackers or pasta, leading to fragility and breakage.

As with other doughs, it is important to control the viscoelastic properties of a cracker dough, especially a saltine cracker dough, to ensure that it can be sheeted and formed to meet the product definition. Saltine crackers are made with a sponge-dough mixing process. Consequently, for this product, in addition to controlling mixing parameters, flour-water relationships, and flour properties, it is important to control the conditions and times of the fermentation. The bacteria associated with the yeast are considered to have the primary role in saltine cracker fermentation. Organic acids produced by the bacteria function as acidulants and cause the liberation of carbon dioxide when they react with the sodium bicarbonate added to the formula. The action of the bacteria also affects the viscoelastic properties of the sponge and the final dough. As fermentation time increases, the dough becomes both less elastic and more extensible because of the effects of pH and flour proteases on the gluten proteins.

Other product quality problems associated with crackers include curling and *checking*. Curling occurs when there is uneven moisture loss from the top and bottom of the cracker. To avoid this, crackers are baked on a wire mesh screen. Checking can be avoided if the crackers are cooled slowly after baking. The cracker's final moisture is very important to texture as well. Crispness, the primary sensory attribute, is optimal at about 2% moisture.

COOKIES

Controlling the dimensions and surface impressions of cookies is the primary issue facing cookie manufacturers. If a cookie is too large in diameter or too thick, it will not fit into the packaging for which it was designed, and if surface impressions are not preserved through baking, many cookies (e.g., animal crackers) lose their identity.

Some cookie formulas are so dry that, in the raw dough, the gluten in the flour does not have sufficient moisture to hydrate or develop. There is also not enough water and usually too many other ingredients (e.g., sucrose) competing for what little water there is for the starch to gelatinize when the dough is baked. The amount a cookie spreads during baking is clearly controlled by the viscosity of the system. The lower the viscosity, the more the cookie spreads. The first event affecting viscosity is the melting of the shortening, which decreases viscosity and initiates spread. As the temperature rises, the solubility of sucrose in the limited water rises as well. When the sucrose enters solution, more fluid volume is created, and the viscosity continues to drop. As the leavening reacts, the cookie expands in all directions. Because it can be shown that starch does not gelatinize in most cookies, the process that increases the viscosity and halts the spread of the cookie is hydration and thermally induced change in the gluten structure.

Thus, cookie spread is controlled by regulating the moisture, the water relationships in the dough, the amount and type of shortening, the pH of the flour (i.e., the extent of chlorination), the amount and solubility of the sucrose included, and the baking conditions. Generally a drier formula with more flour, less shortening, and less sucrose does not spread as much as a richer formula. Because cookie spreading requires time, it may also be possible with some formulations to reduce cookie spread by increasing oven temperatures and reducing bake times.

Upon exiting the oven, most cookies are soft. However, after cooling, many will firm considerably. This is governed by recrystallization of the sucrose in the formula. If a cookie that is by definition firm (e.g., a ginger snap cookie) remains pliable, it is likely that not enough water was driven off during the bake. It is also possible that another type of sugar (e.g., dextrose) was included in the formula and is interfering with the recrystallization of sucrose.

BISCUITS

Many of the problems associated with biscuits are similar to those associated with breads, possible solutions to which were outlined in Chapter 6. Of course, biscuits are chemically leavened, and thus gas generation problems must be addressed by adjusting the levels and/or types of chemical leavening agents. Chemical leavening can also cause appearance issues because Maillard browning is pH-dependent. For example, a biscuit that is too dark under standard baking conditions may have been formulated with excess sodium bicarbonate.

PIE CRUSTS

Most pie crust quality problems relate to texture or, in the case of prepared pie crusts, fragility. For a pie crust to have a flaky texture, it is important for the fat to be dispersed in "islands" and not uniformly throughout the dough during mixing. Then, when the islands are flattened between rollers, they form sheets that separate the flour-water dough portions of the mass, and the baked crust is flaky. Air is also incorporated during this layering process, which also contributes to flakiness in the final crust. General factors controlling the dispersion of fat in the dough are work put into the dough during mixing, consistency of the fat (i.e., hard or soft), and dough processing parameters such as temperature.

Pie crusts are by nature fragile, and breakage of prepared crusts is common. Excessive fragility is usually caused by low moisture in the formula. This is a delicate balance, however, because excess moisture in a pie crust formula produces tough crusts. The shortening or lard can also affect fragility. All fats are partially liquid and partially solid at room temperature. A fat that is too solid can also contribute to a pie crust being brittle and breaking easily.

Web Sites

American Institute of Baking—www.aibonline.org
See description in Ch. 6.

National Baking Center—www.nationalbakingcenter.com
See description in Ch. 6.

The Retailer's Bakery Association—www.rbanet.com
See description in Ch. 6.

Troubleshooting

This section lists some common problems, causes, and suggestions for changes to consider in formulation and processing. However, any flour-based product is complex, and a simple solution is not always possible. Consequently, a solution may involve one or more of these factors or others specific to the system in question. This guide may serve as a starting point for the solution to a problem, but a solution based on fundamental knowledge and/or experience with the product generally leads to a better, longer-lasting resolution to the problem.

CAKES

Symptom	Possible Causes	Changes to Consider
Low volume	Poor gas generation	Incorporate more air in during mixing. Increase chemical leavening.
	Poor gas retention	Increase batter viscosity (e.g., add flour, remove water, add egg whites, etc.). Decrease batter hold time. Avoid vibrations. Check pH.
	Structure setting before or after maximum expansion	Adjust sucrose, water, and flour levels.
Coarse texture	Poor air bubble dispersion	Add or adjust surface-active agents. Increase batter viscosity (e.g., add flour, remove water, etc.).
Gum layer	Starch settling	Increase batter viscosity (e.g., add flour, remove water, etc.).
Flavor	pH not optimal	Adjust leavening agents. Check flour pH.
Gluten separation	Overmixing of batter	Decrease work input. Use lower-protein flour. Add starch or remove flour.

DOUGHNUTS

Many issues affecting doughnuts are similar to those affecting cakes, and the cake troubleshooting guide may apply. Changes to consider that are specific to doughnuts are as follows:

Symptom	Possible Causes	Changes to Consider
Greasiness	Excess fat pickup	Reduce formula moisture.
Toughness	Overdevelopment	Reduce mix time.
Fragility	Underdevelopment	Increase mix time.

CRACKERS

Symptom	Possible Causes	Changes to Consider
Dough difficult to sheet	Non-optimal viscoelastic properties	Adjust fermentation time and conditions (saltine crackers). Adjust mixing time. Adjust flour water levels. Check to assure flour meets specification.
Curling	Uneven moisture loss during baking	Assure uniform oven temperature. Bake on mesh screens.
Checking	Rapid cooling	Reduce rate of cooling.
Toughness	High moisture	Adjust baking conditions to lower moisture content.

COOKIES

Symptom	Possible Causes	Changes to Consider
Excess spread/surface distortions	Low viscosity during bake	Add flour. Remove water, sugar, and/or shortening. Reduce bake time and raise bake temperature.
	Excess gas production	Reduce leavening.
Wafer sticks to plates	Surface irregularities	Apply non-stick coating to plates.
Finished cookie too soft	Sucrose crystallization inhibited	Increase bake time or temperature. Remove other sugars from the formula.

BISCUITS

Since biscuits are very similar to breads, many issues are common and can be resolved with similar changes to the process or the product formulation. The troubleshooting guide for breads in Chapter 6 may be a useful reference. If an issue cannot be resolved with the changes suggested there, a change in the level or type of chemical leavening agents should be considered.

PIE CRUSTS

Symptom	Possible Causes	Changes to Consider
Toughness	Non-optimal distribution of shortening or lard	Reduce mixing time. Raise dough temperature during mix. Use harder shortening or lard. Reduce water content.
Fragility	Inadequate development	Increase water content. Use softer shortening or lard.

References

1. Hoseney, R. C. 1994. *Principles of Cereal Science and Technology*, 2nd ed. American Association of Cereal Chemists, St. Paul, MN. Chapter 13.
2. Ghiasi, K. Hoseney, R. C., and Varriano-Marston, E. 1983. Effects of flour components and dough ingredients on starch gelatinization. Cereal Chem. 60:58-61.
3. Hoseney, R. C., Wade, P., and Finley, J. W. 1988. Soft wheat products. Pages 407-456 in: *Wheat: Chemistry and Technology*, 3rd ed., Vol. 2. Y. Pomeranz, Ed. American Association of Cereal Chemists, St. Paul, MN.

Supplemental Reading

1. Pyler, E. J. 1988. *Baking Science and Technology*, 3rd ed., Vol. 1. Sosland Publishing Co., Merriam, KS. Chapters 24 and 25.
2. Faridi, H. 1994. *The Science of Cookie and Cracker Production*. Chapman and Hall, New York.

CHAPTER 8

Durum-Based Products: Problems, Causes, and Resolutions

In the United States, durum wheat is used almost exclusively to make pasta products. These products are extremely popular all over the world and come in a variety of forms (Fig. 8-1). Spaghetti, lasagna, linguine, and vermicelli are examples of pasta products known as "long goods." Macaroni, rigatoni, ziti, and penne are examples of "short goods." All of these products are identical in formulation, although they have many shapes and sizes. The formulas used to produce pasta are simple, usually consisting of only semolina, water, and possibly enrichment.

Noodles are related products but are often made with soft wheat flour instead of semolina. Common ingredients found in noodles, in addition to semolina or flour and water, are salt and eggs. Although these products are not often based on durum wheat, they are discussed here because of their similarities to pasta in formulation, processing, and the issues associated with them.

In This Chapter:

Ingredients and Formulation
- Pasta
- Noodles

Processing
- Pasta
- Noodles

Product and Processing Issues
- Pasta
- Noodles

Troubleshooting

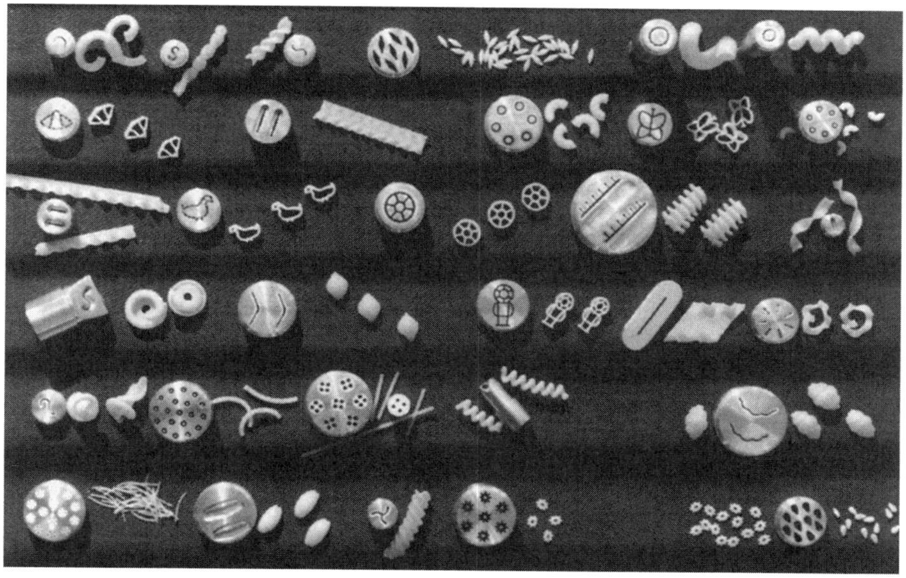

Fig. 8-1. Various pasta shapes. (Reprinted from [1])

Ingredients and Formulation

PASTA

Semolina from durum wheat is preferred for the production of pasta products. Pasta made from semolina has a desirable yellow color in the dry form and is durable enough for packaging and shipping without a high degree of breakage. When cooked, the pasta is cohesive, and it does not break apart or split while being boiled in water. The cooking water is free of starch, and the product resists overcooking. These characteristics in the pasta product are provided by semolina as a result of its hardness, the characteristics of its gluten protein matrix, and the dispersion of the pigments throughout the endosperm. Other ingredients often included in pasta are spinach and tomatoes. According to the U.S. Code of Federal Regulations (21 CFR 139), the only other ingredient required to make pasta is water.

NOODLES

Noodles are made with semolina and all types of flour, but soft, white wheat flour is often preferred. If strong, high-protein flour is used, the noodles are too elastic and chewy when cooked. A weak, low-protein flour, however, is also not desirable because the noodles become very sticky or disintegrate when cooked. Since appearance is important, the flour used in noodle production should also be free of bran, especially if a red wheat is used. Consequently, low-extraction flours are often used. It is also important to ensure that enzyme activity in the flour is low. For example, if sprout-damaged flour is used, the α-amylase contained in it causes a rapid breakdown in the integrity of the noodle. Polyphenol oxidase, which fosters the production of dark pigments, is also detrimental to quality.

Noodles for Asian markets are made from flour with very strict quality requirements. Japanese noodles require a low-ash (0.36–0.40%), low-protein (8–10%) wheat flour that has a creamy white appearance. The noodle produced from Australian standard white (ASW) wheat is preferred because of its good surface appearance, favorable texture, and minimal cooking loss. Amylograph curves of ASW wheat flour are similar to those of native Japanese wheats, and it is possible that this characteristic is important to the final quality of the noodle. Chinese noodles are made from hard wheat flours with very low ash content (i.e., 0.33–0.38%) and protein amounts ranging from 10.5 to 12%. Color and texture are very important quality attributes, and consequently, the flour must deliver these characteristics.

Noodles, like pasta, are made from very simple formulas. Noodles made in the United States include eggs, but in other parts of the world, egg is not a required or desired ingredient. Salt is used in noodle formulations everywhere. In addition to flavor, salt affects the texture of the noodle, generally making it more cohesive and elastic. Some noodles are fried, and therefore oil, absorbed during frying, is a component of these noodles.

Processing

PASTA

Pasta production consists of three basic operations: mixing, forming, and drying (Fig. 8-2). Pasta dough is usually formulated at about 31% moisture. This is a low moisture level compared with those of bread doughs, which generally are about 45%. When mixed, pasta dough generally has the appearance of wet sand, with dough balls of about 2.5 cm (1 in.) in diameter. The dough is mixed in an airtight paddle mixer. Air is excluded during mixing by the creation of a partial vacuum to ensure that the final product does not contain microscopic air bubbles. Air diffuses light; hence, air bubbles cause the pasta to appear opaque and detract from the translucent appearance desirable in pasta products. Additionally, air bubbles in dried pasta are points of weakness that foster breakage. The oxygen in air is also required in the bleaching reaction that affects durum pigments and is catalyzed by the enzyme lipoxygenase. Hence, minimizing air in the pasta also inhibits oxidation of the desirable yellow pigments.

After mixing, the dough is fed into the barrel of a cold forming extruder, where it is subjected to high pressure as it transverses the length of the barrel from the infeed to the die. The considerable friction involved in extrusion generates heat, which can be detrimental.

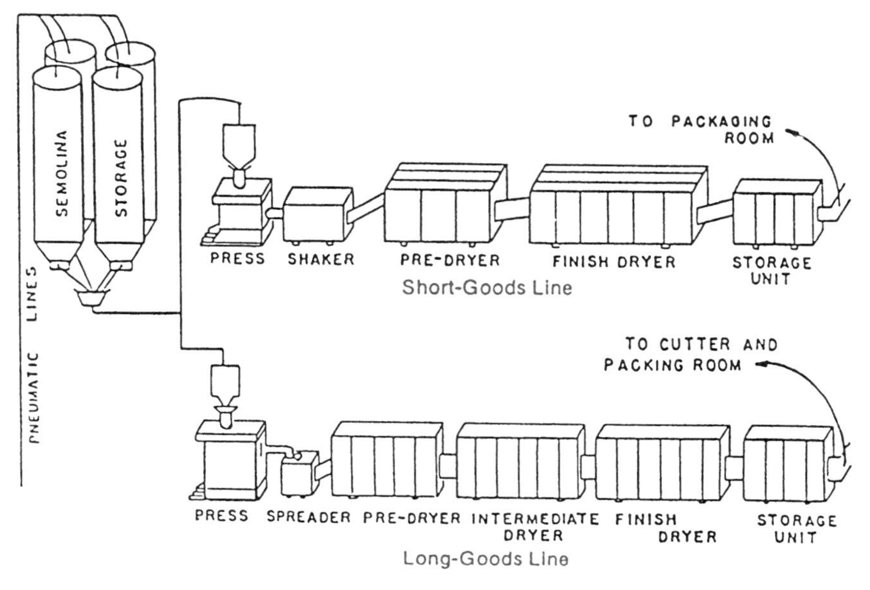

Fig. 8-2. Production line schematic for long- and short-goods pasta. (Reprinted from [2])

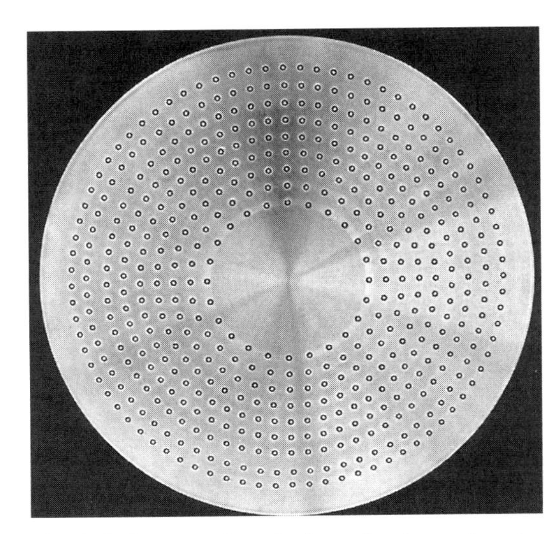

Fig. 8-3. An extrusion die for a short-good pasta. (Reprinted from [1])

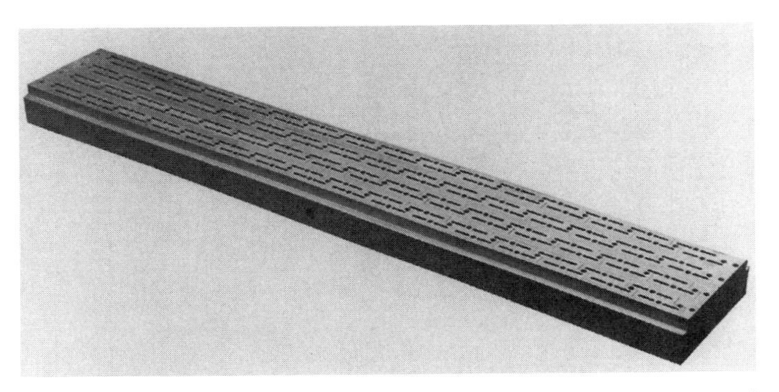

Fig. 8-4. An extrusion die for a long-goods pasta. (Reprinted from [1])

Therefore, pasta extruders are usually jacketed with cold water to avoid excessive temperatures.

Dies are available in many shapes and sizes representative of the variety of pasta products made (Figs. 8-3 and 8-4). The surface of the die can affect the cooking quality of the pasta because it molds the surface of the extruding pasta dough. If the pasta surface is irregular, there is more surface area, which causes the product to cook faster. If the surface has major flaws, parts of the pasta may break off during cooking. A surface that is too smooth hydrates and cooks more slowly and often yields a sticky or mushy exterior. Bronze dies produce the best product, but because such dies lack durability, stainless steel and Teflon-coated dies are often used.

After extrusion, the product is cut to the desired length. Long goods are draped over rods for transport through the drying process. Short goods are shaken to avoid clumping and are transported through the dryers on conveyor belts.

The dough exiting the die contains about 31% moisture, and the finished pasta contains about 12%. The drying operation must be engineered to accomplish this slowly. Rapid drying leads to checking (i.e., fissures in the product), which is a major cause of breakage. If the product is dried too slowly, however, it can become moldy, or in the case of long goods, it can stretch and become misshapen. Drying is generally accomplished in three phases. The first is a rapid drying phase that dehydrates the surface. Although this is only about 10% of the drying time, it removes about one-third of the moisture. It is followed by a "sweating" period of 2–4 hr, which allows the water in the product to slowly equilibrate at 90% relative humidity. The final phase is a long, slow drying to 12% moisture that usually takes 10–16 hr. Modifications of this process may include several alternate periods of drying and sweating.

NOODLES

Noodle processing generally involves sheeting rather than extrusion. Doughs are mixed at about 35% moisture. Because this is not enough water to develop the gluten to the extent that it is developed in bread products, the primary purpose of the mixing is to hydrate the flour and other ingredients. After mixing, the dough is allowed a rest period to fully hydrate the flour particles. The dry, noncohesive dough is then fed through a series of sheeting rolls and sheeted to a final thickness of about 2 mm. Sheeting is unidirectional, and the gluten fibrils are therefore aligned to give the product the most strength in the lengthwise direction. The thin sheet is then cut in the long direction, followed by another cut to the desired length. If the noodle is to be dried, the drying process is usually similar to the process described above for pasta. In many parts of the world, noodles can also be fried or sold fresh. In the case of Oriental noodles, the many different types (Fig. 8-5) result from a variety of related but unique processing schemes (Fig. 8-6).

Fig. 8-5. Various types of oriental noodles. Top, dried noodles. Bottom left, wet noodles; center two, ramen noodles; right two, packaged wet noodles. (Courtesy P. A. Seib)

Product and Processing Issues

PASTA

There are two general causes of problems associated with pasta products: those involving semolina and those involving processing practices. Good uniform color is a very important quality characteristic, and therefore it is important that the semolina purchased be a bright yellow color. The pigments of wheat are susceptible to oxidation, so it is very important that additives such as benzoyl peroxide and chlorine be strictly avoided. Oxidation can also be catalyzed by the enzyme lipoxygenase. Pasta made with semolina containing high levels of lipoxygenase is not yellow; therefore, a semolina lipoxygenase assay may be warranted if color is an issue.

If raw pasta is opaque and not translucent, it has air incorporated

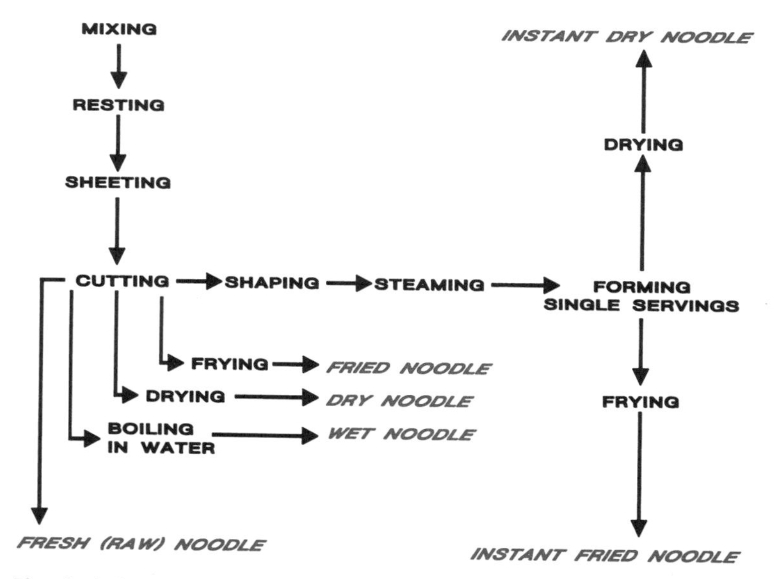

Fig. 8-6. Processing of various types of oriental noodles. (Reprinted from [3])

Blackpoint—An infection of wheat kernels by the fungi *Alternaria alternata* and *Helminthosporium sativum.*

into it. Translucency occurs because a material is uniform and is uninterrupted by boundaries of materials that bend light differently. If air is present, many boundaries exist between the pasta and air bubbles, and light is bent many times as it transverses the product. Air in pasta results from the mixing process; as with soft and hard wheat products, air is incorporated by the mixer. Gas retention in pasta dough is poor, but retention of even relatively few bubbles can diffuse light to the extent that the pasta has an opaque appearance.

Specks in pasta are undesirable. One cause is the inclusion of bran in the semolina. Reducing the allowable ash in the semolina specification may be necessary if this is an issue. Specks can also be caused by infection of wheat kernels by the fungi *Alternaria alternata* and *Helminthosporium sativum*. This condition, called *blackpoint*, can discolor the endosperm. Reducing ash content cannot eliminate this cause of specks. Instead, the miller must employ rigorous selection criteria to ensure that kernels affected by blackpoint are not used in the preparation of semolina. Another cause of specks in pasta and other products can be the enrichment. The iron in enrichment can foster the polymerization of dark pigments similar to those involved in the polyphenol oxidase reaction. If this is the cause of a problem, ensuring that there are no clumps in the enrichment may help. It may also help to use a less reactive form of iron in the enrichment (e.g., reduced iron instead of ferrous sulfate).

Moisture control during processing is critical. If a dough is too wet, the pasta is extruded too fast. Control of shapes may be difficult because the wet pasta stretches easily, which causes nonuniform thickness and diameter upon further processing. Wet doughs also require prolonged drying times. Conversely, a dry dough is not cohesive coming out of the extruder, and the fresh and dry pasta may be fragile as a result. The material may also be difficult to mold, which can cause the product to be misshapen.

Perhaps the most common problem associated with pasta products is checking. When hygroscopic materials are wet, they swell, and when they dry, they contract. This is also the case with the starch, gluten, and pentosans, which make up the major portion of a pasta product. If sufficient time is not allowed for the water to equilibrate across the product, nonuniform contraction causes small fissures to form. In many manufacturing settings, speed of production is an important economic factor. Unfortunately, there are no alternatives to time when drying high-quality pasta.

NOODLES

Maintaining good flour quality is essential to the production of high-quality noodles. Flour specifications should contain criteria for wheat type, protein level, protein quality (e.g., farinograph characteristics), and ash. Soft wheat flour is generally used for noodle production. The gluten in it must be of sufficient strength to produce a noodle that can be cooked without disintegrating or becoming sticky. It must also cook to produce a noodle that is not too elastic and therefore perceived as chewy. Controlling protein level and farinograph characteristics within predetermined levels is the best way to ensure that optimal noodle strength is maintained, although processing parameters can influence strength as well. Optimal moisture relationships and hence dough viscoelastic properties should be identified and maintained, and sheeting should always be unidirectional to align the gluten fibrils. Since the sheet is strongest with the direction of sheeting, the noodles should always be cut so that the gluten fibrils are aligned lengthwise (parallel to the long axis of the noodle).

Flour quality is also important to the appearance of noodles. Low-extraction flours produce noodles with the fewest specks. A severe problem can occur if polyphenol oxidase activity is high in the flour. This causes the noodles to turn gray and appear very unpalatable. Minimizing the ash level usually solves both of these problems because polyphenol oxidase is located in the aleurone, which is closely associated with bran. As stated in the pasta section above, blackpoint and enrichment can also be causes of specks.

Checking is generally not as likely to occur with noodles as it is with pasta products. The salt in the noodle formulas retains moisture and thus slows the rate of dehydration. Hence, the rapid dehydration that causes nonuniform contraction generally does not occur. If checking is an issue, however, more equilibration time in the drying process is required.

Web Sites

U.S. Durum Growers Association—www.durumgrowers.com

The U.S. Durum Growers Association was organized to gain identity recognition for durum, encourage greater consumption of durum and durum-related products, and increase the production potential of durum. The Association also supports agronomic research and promotes durum and durum-related products.

National Pasta Association—www.ilovepasta.org

The National Pasta Association is the trade association for U.S. pasta manufacturers and related suppliers.

Troubleshooting

This section lists some common problems, causes, and suggestions for changes to consider in formulation and processing. However, any flour-based product is complex, and a simple solution is not always possible. Consequently, a solution may involve one or more of these factors or others specific to the system in question. This guide may serve as a starting point for the solution to a problem, but a solution based on fundamental knowledge and/or experience with the product generally leads to a better, longer-lasting resolution to the problem.

PASTA		
Symptom	**Possible Causes**	**Changes to Consider**
Fragility	Checking due to nonuniform contraction	Allow more equilibration time during drying.
	Air bubbles	Eliminate air in the mixing process.
	Dough too dry	Add water.
Misshapen product	Dough too wet	Remove water.
Opacity	Air bubbles	Eliminate air in the mixing process.
Specks	Bran	Lower ash level of semolina.
	Blackpoint	Inform miller.
	Enrichment clumps	Reduce enrichment particle size.
Undesirable color	Semolina color	Ensure that semolina color is within specification.
	Lipoxygenase	Reduce semolina ash level. Assay lipoxygenase.
	Oxidizing agent	Ensure that there is no benzoyl peroxide or similar additive in semolina.

NOODLES		
Symptom	**Possible Causes**	**Changes to Consider**
Stickiness with poor cook resistance or chewiness with high cook resistance	Nonoptimal gluten development	Adjust protein level of flour. Adjust farinograph parameters of flour. Adjust flour-water relationship in dough.
Fragility	Nonoptimal dough	Adjust flour-water moisture relationships in dough.
	Poor gluten alignment	Ensure that dough is sheeted unidirectionally and that the long axis of the noodles is in the same direction as the sheeting.
Graying	Polyphenol oxidase	Lower ash content of flour.
Specks	Bran particles	Lower ash content of flour.
	Blackpoint	Inform miller.
	Enrichment clumps	Reduce enrichment particle size.
Checking	Nonuniform contraction	Allow more equilibration time during drying.

References

1. Dalbon, G., Grivon, D, and Ambrogina Pagani, M. 1996. Continuous manufacturing process. Pages 13-58 in: *Pasta and Noodle Technology*. J. E. Kruger, R. B. Matsuo, and J. W. Dick, Eds. American Association of Cereal Chemists, St. Paul, MN.
2. Dick, J. W., and Matsuo, R. R. 1988. Durum wheat and pasta products. Pages 507-547 in: *Wheat: Chemistry and Technology*, 3rd ed., Vol. 2. Y. Pomeranz, Ed. American Association of Cereal Chemists, St. Paul, MN.
3. Oh, N. H., Seib, P. A., Deyoe, C. W., and Ward, A. B. 1983. Noodles. I. Measuring the textural characteristics of cooked noodles. Cereal Chem. 60:433-438.

Supplemental Reading

1. Kruger, J. E., Matsuo, R. B. , and Dick, J. W. 1996. *Pasta and Noodle Technology*. American Association of Cereal Chemists, St. Paul, MN.
2. Fabriani, G., and Lintas, C. 1988. *Durum Wheat: Chemistry and Technology*. American Association of Cereal Chemists, St. Paul, MN.
3. Hoseney, R. C. 1994. *Principles of Cereal Science and Technology*, 2nd ed. American Association of Cereal Chemists, St. Paul, MN. Chapter 15.

Glossary

Acidulant—The acidic portion of a chemical leavening system. The acidulant produces acid, which reacts with bicarbonate to produce carbon dioxide.

Albumins—Proteins that are soluble in water. Many enzymes are albumins.

Aleurone—The outermost layer of endosperm cells, biologically active and with a high enzyme content.

Amino acids—The building blocks of proteins. Each contains an amine group, a carboxylic acid group, and a side group. The amine and acid groups form peptide bonds with other amino acids to form protein chains. The side groups influence how the protein interacts with other molecules.

α-Amylase—An enzyme that severs the α-1,4 bonds between glucose units in starch. Sprouted wheat has high α-amylase activity.

β-Amylase—An enzyme that hydrolyzes α-1,4 bonds in amylose and amylopectin. It successively removes maltose from the nonreducing ends of starch polymers but cannot cross a branch point.

Amylopectin—A branched polysaccharide composed entirely of glucose units. It is one of the two major components of starch.

Amylose—A linear polysaccharide composed entirely of glucose units. It is one of the two major components of starch.

Arabinose—A five-carbon sugar that composes the side chains of arabinoxylans.

Arabinoxylan—A nonstarchy polysaccharide that is a major component of the cell walls of wheat.

Ash—Material, composed primarily of minerals, surviving very high temperature treatment of flour or wheat.

Aspiration—The process of using circulating air to separate materials of different densities. Low-density materials are levitated by a current of air while denser materials remain stationary.

Azeotrope—A miscible mixture of solvents that cannot be separated by boiling. Ethanol and water form an azeotrope at 95% ethanol that boils at 78.2°C (173°F).

Baking powder—A preparation containing sodium bicarbonate and one or more acidulants.

Birefringence—The phenomenon that occurs when polarized light interacts with a highly ordered structure, such as a crystal. A crossed diffraction pattern, often referred to as a "Maltese cross," is created by the rotation of polarized light by a crystal or highly ordered region, such as that found in starch granules.

Blackpoint—An infection of wheat kernels by the fungi *Alternaria alternata* and *Helminthosporium sativum*. This condition can discolor the endosperm and cause specks in pasta and noodle products.

Bleaching agent—A chemical added to flour to bleach pigments and thus whiten flour. Benzoyl peroxide is a common bleaching agent.

Boot—A swelling in the wheat stem caused by the developing wheat head. The head is in the boot just before emergence.

Bran—The outer protective layers of a wheat kernel. Bran can be pigmented, as in the red wheats, and is high in fiber content.

Break system—The initial process in wheat grinding. Wheat is fed through corrugated rolls in the first break and subsequently ground finer in the corrugated rolls of later breaks.

Caramelization—The reaction occurring when sugars dehydrate and polymerize at high temperatures, producing brown pigments and flavor compounds that contribute to the flavor profile of many wheat-based products.

Carbohydrates—Substances (e.g., sugars and polysaccharides) that generally conform to the molecular structure $C_X H_{2x} O_X$. Sugars (e.g., glucose, maltose, and sucrose) and polysaccharides (e.g., amylose and arabinoxylan) are carbohydrates.

Carotenoids—The yellow pigments found in grains, including wheat. They are especially prevalent in durum wheat and give pasta its characteristic yellow color.

Celiac disease—A chronic abdominal disease caused by allergy to one of the protein fractions of wheat.

Checking—The production of fissures in crackers or pasta, leading to fragility and breakage.

Chlorination—The process of applying gaseous chlorine to flour. Chlorination lowers the pH of the flour and improves the cake-making ability of soft wheat flour.

Chlorine—A gas often applied to soft wheat flour to improve cake-making ability. The pH of chlorinated flour is lower than that of its untreated counterpart.

Chromosome—A body composed of DNA and carrying part of the genetic code for the organism, i.e., the wheat plant. There are seven distinctly different chromosomes in wheat.

Class—A type of wheat usually designated by hardness, color, and growing season. Hard red spring wheat is a class of wheat that is commonly used to produce bread. Durum is a class of wheat used to make pasta products.

Cleaning—The process of removing unwanted material (i.e., dockage) from wheat before tempering and milling.

Clear flour—A high-ash flour fraction consisting of the portion of flour between 65 and 72% extraction. It is also called low-grade flour.

Conditioning—A process in which the equilibration of wheat to a higher moisture level during tempering is facilitated by heat.

Cutoff flour—The portion of flour between 45 and 65% extraction.

Cysteine—An amino acid with a sulfur-containing side group that can bind to other cysteine groups to form disulfide linkages.

Damaged starch—Starch granules that have been physically broken during milling.

Denaturation—The process of destroying the tertiary structure of a protein. Denaturing an enzyme eliminates its activity as a catalyst.

Dextrins—Small starch fragments (linear or branched) formed when starch bonds are severed.

Diploid—Describing an organism with two sets of chromosomes. An early wheat known as einkorn was diploid.

Disk separator—A wheat-cleaning machine having cavities in rotating disks that exclude or accept grains based on size.

Disulfide bonds—"Bridges" between cysteine residues in the same protein chain or in adjacent protein chains.

Dockage—Unwanted material in wheat coming into a mill, principally insects, stones, straw, and other contaminants.

Dough development time—A farinogram parameter, also called mixing time or peak time, which gives the time between the origin of the curve and its maximum. Monitoring this parameter aids in making mixing time adjustments in commercial processes when the flour mixing requirements change.

Dry stoner—A machine that utilizes air to separates materials of different densities. It aspirates the wheat while not aspirating heavier materials such as stones.

Emulsifier—An additive that aids the dispersion of fat into a formula. Monoglycerides and diglycerides are common emulsifiers.

Encapsulation—A technology that enables small particles or droplets of materials to be coated, usually with a fat. Encapsulated sodium bicarbonate reacts more slowly than free sodium bicarbonate.

Endosperm—The major portion of the wheat kernel by weight. It is the primary constituent of flour and contains starch and gluten.

Enrichment—Nutrients added to flour for nutritional purposes. Bread enrichment consists of B vitamins (i.e., niacin, thiamine, riboflavin, and folic acid) and iron.

Enzymes—Proteins that function as catalysts.

Extraction rate—The amount of flour made as a percentage of total wheat ground. Whole wheat has an extraction rate of 100%. A straight-grade flour, which has an extraction rate of 72%, contains much less bran and germ.

Falling number—The time it takes (in seconds) for the viscometer stirring rod in a falling number apparatus to fall through the starch paste. High falling numbers indicate flours with low α-amylase activity.

Farinograph absorption—The amount of water added to balance the farinograph curve on the 500-BU line, expressed as a percentage of the flour (14% mb). This parameter is useful in adjusting the water relationships in commercial doughs when flour changes.

Fiber—Carbohydrates that cannot be digested in the human gut. Pentosans and β-glucans are fibers found in wheat.

Flour dressing system—The final sieving of flour through 10XX bolting cloth (i.e., a sieving material with apertures of 136 μm).

Foam—A gaseous noncontinuous phase held in a continuous phase. An aerated batter or dough is an example of a foam.

Futures contract—An agreement between a seller and buyer in which a price is negotiated in advance for wheat that will be delivered at some point in the future.

Gelatinization—Irreversible loss of the molecular order of starch granules, shown by swelling and loss of crystallinity. Heating hydrated native wheat starch causes it to gelatinize.

Germ—The potential wheat plant within the wheat kernel. It is high in oil content and many nutrients.

Germination—The process whereby the wheat seed begins to grow. Warm temperatures and moisture initiate germination.

Gliadin—One of the two major components of gluten. It is composed of individual protein chains and imparts the viscous nature to gluten.

Globulins—Proteins that are soluble in salt solutions. Many enzymes are globulins.

β-Glucans—Nonstarchy polysaccharides composed entirely of glucose arranged linearly and bound together with β-1,3 and β-1,4 bonds. They are a component of wheat cell walls.

Glucose—A six-carbon sugar that is the "building block" of amylose, amylopectin, and β-glucan.

Glutamine—An amino acid that makes up about 40% of the gluten protein. The side group of glutamine is an amide, which binds water well.

Glutelins—Proteins that are soluble in dilute acids.

Gluten—The primary protein complex found in wheat. In many applications, gluten content and quality play a major role in how the flour functions in an end product.

Glutenin—One of the two major components of gluten. It is a very large molecule that consists of protein chain subunits connected by disulfide linkages.

Gravity table—A vibrating inclined plane that cleans wheat by separating materials based on density.

Heading—A stage in the growth cycle of a wheat plant initiated when the wheat head emerges from the boot. Heading is complete when the head has flowered and the seeds begin to develop.

Hedging—A procedure using futures contracts to minimize economic risks caused by fluctuations in wheat market prices.

Hemicellulose—A term often used interchangeably with "pentosan" to describe the nonstarchy polysaccharides of flour.

Hexaploid—Describing an organism with six sets of chromosomes. Most common wheat is hexaploid.

Hydrolysis—A chemical process in which a peptide or glycosidic bond is severed after the addition of water.

Identity preservation—The segregation of wheat varieties to allow them to remain uncontaminated with other varieties between production and end-product use.

Jointing—The first portion of the stem-extension phase of wheat growth. During jointing, two observable joints (i.e., nodes) develop in the stem.

Kernel—An individual seed of a cereal grain.

Keyholing—Contraction and collapse of the side walls of a loaf of bread upon cooling.

Lard—Solid fat derived from pigs. It is preferred for the production of pie crusts.

Lipase—An enzyme that catalyzes the reaction that severs bonds between fatty acids and glycerol in triglycerides.

Lipids—The class of compounds that includes fats and oils. Examples of flour lipids are triglycerides, carotenoids, and tocopherols.

Lipoxygenase—An enzyme that catalyzes the reaction between oxygen and double bonds in fatty acids and other lipids, often leading to the development of rancidity.

Lysis—Rupture of the cell walls of a microorganism. Yeasts lyse in doughs if they are overfermented.

Magnetic separator—Devices to remove tramp metal during wheat cleaning. They do not remove metals that are not magnetic.

Malt—Barley that has been allowed to germinate. It is ground and added to flour to bolster enzymatic activity and/or to improve flavor.

Maltose—A sugar composed of two glucose molecules bound by α-1,4 bonds. It is liberated by the action of β-amylase on amylose or amylopectin.

Middlings—Large particles of endosperm obtained after the break system in a mill. They are reduced to flour in the reduction system.

Millfeed system—The process in a flour mill that separates and purifies flour milling by-products, specifically bran and germ.

Milling—Grinding of wheat. In a larger sense, all aspects of the conversion of wheat to flour, including cleaning, tempering, grinding, sieving, purifying, etc.

Milling separator—A machine that cleans wheat by drawing lighter, less-dense materials away from wheat in a current of air.

Oxidants—Compounds added to dough to facilitate the formation of disulfide bonds. Oxidants have a strengthening effect on dough and make them more elastic and less extensible.

Pasting—The breaking down of the starch granule following gelatinization.

Patent flour—Straight-grade flour with some of the higher-ash components removed. Extraction rates range from 45% (i.e., short-patent flour) to 65% (i.e., long-patent flour).

Pentosanase—An enzyme that catalyzes the reaction that severs bonds in arabinoxylans. Many specific types of pentosanases exist.

Pentosans—A group of polysaccharides containing five-carbon sugars and constituting a major portion of the nonstarchy polysaccharides of wheat. Arabinoxylans are pentosans.

Peptide bonds—The type of bond that forms when the amine group of an amino acid reacts with the carboxylic acid group of another amino acid.

Phytase—An enzyme that catalyzes the reaction that severs bonds between the phosphate groups and inositol in phytate.

Phytate—A molecule consisting of phosphate groups bound to the vitamin inositol. Phytate binds calcium and magnesium and is undigestible.

Pistil—The female sex organ of plants. During the heading phase of wheat development, the pistils are pollinated to initiate the development of new wheat kernels.

Plasticizer—An ingredient that renders a mixture more flowable or pliable.

Polyphenol oxidase—An enzyme that catalyzes a reaction between oxygen and the phenolic compounds in flour. The polymers that result are responsible for the darkening of dough observed when it is exposed to air.

Polysaccharides—Polymers composed of sugar units.

Primary structure (of a protein)—Its sequence of amino acids.

Prolamins—Proteins that are soluble in aqueous alcohol. Gliadin is a prolamin.

Proline—A cyclic amino acid that creates bends in protein chains.

Proteases—Enzymes that catalyze the reaction that severs peptide bonds in proteins. There are many types of proteases in wheat, but their combined activity is usually low.

Purification system—The part of a durum mill after the break system in which rough semolina is gradually reduced in size. The rollers in the roller mills are corrugated but designed to minimize the production of small particles.

Purifiers—Machines that remove light, low-density material from higher-density particles using aspiration. They usually follow roller mills in a milling process.

Reducing agents—Chemicals added to dough to inhibit the formation of disulfide linkages. Doughs formulated with reducing agents are more extensible and less elastic than untreated doughs.

Reduction system—A milling process that employs roller mills with smooth rolls to reduce particle size. The reduction system in a hard or soft wheat mill is longer than that in a durum mill.

Retrogradation—The process in which dispersed starch polymers recombine to form crystal structures.

Ripening—The final stage in the growth cycle of wheat, in which seeds mature. At the beginning of ripening, the seeds are soft and have a "milky" texture. At maturity, the kernels are hard and ready to harvest.

Roller mill—The machine in a mill that grinds wheat, as well as wheat particles from roller mills earlier in the process. Sifters and purifiers that separate the ground components always follow roller mills.

Scourer—A wheat-cleaning machine that removes molds and dirt adhering to wheat kernels using an abrasive screen or surface.

Secondary structures (of a protein)—Structures involving single chains. For example, helices and pleated sheets are secondary structures of proteins.

Semolina—The primary product of a durum mill, used almost exclusively to make pasta. Its particles are similar to middlings in size.

Shortening—Solid fat derived from plants. Common sources of shortening used in baked products include soybeans and canola.

Shorts—A very-high-ash product produced in a mill. Shorts are composed primarily of finely ground bran.

Sieves—Devices to separate particles based on size. Smaller particles pass through apertures in the sieve while larger particles are retained. Several sieves are usually used in a series to separate particles of many different sizes.

Sourcing—The process of identifying and procuring a specific type of wheat or flour. It involves the consideration of quality, logistics, and cost.

Species—A biological classification below *genus* and above *variety*. Wheat varieties from three species of wheat are commonly grown.

Specific volume—Volume in cubic centimeters divided by weight in grams.

Sprouted wheat—Wheat that has germinated. Kernels of sprouted wheat have high enzymatic activity and can be unsuitable for production of certain end products.

Stability—A farinograph parameter, defined as the difference in minutes between the arrival time and the time at which the top of the curve falls below the 500-BU line (i.e., the departure time). It is a measurement of how well a flour resists overmixing.

Stamen—The male sex organ of plants. Wheat heads contain multiple stamens and pistils and, consequently, wheat is self-pollinating.

Starch—The primary carbohydrate in the wheat endosperm, composed entirely of glucose units linked together. Starch exists in aggregates known as granules.

Starch granules—Discrete, partially crystalline aggregates of starch in the wheat endosperm, composed of amylose and amylopectin.

Stem extension—The phase of wheat growth following tillering, in which the stem elongates and the wheat head begins to develop within the stem.

Stirring number—The apparent viscosity in Rapid Visco Analyser units after 3 min of stirring. As α-amylase activity rises, the stirring number increases.

Straight-grade flour—The primary product of most flour mills, having an extraction rate of about 72%. It thus contains more non-endosperm components than patent flours.

Strain hardening—Becoming more resistant to extension and harder as a result of being stretched.

Strong flour—Flour that produces a very elastic dough that retains gas well.

Surfactants—Surface-active agents that affect how two materials with different properties interact at the surface between them. Propylene glycol monostearate is a surfactant that can affect how bubbles in a batter coalesce.

Tempering—The process in which water is added to wheat and allowed to equilibrate to toughen the bran and soften the endosperm. It follows cleaning and precedes milling or grinding in a flour mill.

Tertiary structure (of a protein)—The way the entire molecule is oriented in space. The tertiary structure of an enzyme molecule is critical to its catalytic activity.

Tetraploid—Describing an organism with four sets of chromosomes. Durum wheat is tetraploid.

Tillering—The first stage in the development of the wheat plant, initiated when the shoot breaks the ground. During tillering, shoots (i.e., tillers) are sent out laterally to form new wheat plants.

Tocopherol—Lipids found in flour that have antioxidant effects.

Tolerance index—A farinogram parameter(also called mixing tolerance index), measured as the difference in Brabender units between the top of the curve at the optimum and the point on the curve 5 min later. It is similar to stability because it is a measurement of how well a flour resists overmixing.

Triglyceride—Three fatty acids bound to a glycerol molecule.

Variety—A biological classification below *species*. Many varieties of wheat are grown throughout the world, and new varieties are continually being developed.

Vernalization—A process required for a winter wheat to create wheat heads. The temperature must drop below freezing for this to occur. If winter wheat is not vernalized, the plant will grow but never produce heads with new seeds.

Viscoelastic—Describing a substance that has both viscous (i.e., flow) and elastic properties. Gluten is viscoelastic.

Vital wheat gluten—Gluten that is extracted and dried during wet milling. It is often added to weak flours to bolster gas retention.

Weak flour, flour that produces a less cohesive dough that does not retain gas well.

Xylose—The five-carbon sugar that forms the backbone of arabinoxylans.

Yeast food—A minor ingredient composed of nutrients that enhance yeast activity. Prominent among these are ammonium salts.

Yield—The amount of wheat produced per area of land. Common units of yield are bushels per acre and metric tons per hectare.

Index

Absorption
  farinograph, 53, 54
  mixograph, 56
Acidulants, 82, 98, 99
  neutralization value, 99
Additives, legally permitted, 81
Aeration, of batter, 103, 104
Agtron color test, 48–49
Air bubbles, 107, 108
  in pasta, 117, 120
Air cells, in dough, 91, 92, 93
Albumins, 27, 30, 31
Aleurone layer, of wheat kernel, 6, 34, 50–51
Alkaline water retention capacity, 58
Alternaria alternata, 120
Alveograph test, 58, 72
American Association of Cereal Chemists
  analytical methods of, 47
  Web site, 44, 64
American Institute of Baking, Web site, 92, 111
Amino acids, and protein structure, 28, 29, 30–31
Ammonium bicarbonate, 98
α-Amylase
  activity in flour, 38–39
  in bread, 43, 89
  and damaged starch, 63
  in germinating wheat, 7
  from sources other than wheat, 39
  tests for, 59–61, 71
β-Amylase, 39, 43
Amylograph test, 60, 61, 72
Amylopectin
  content in waxy wheat, 3
  and crumb firming, 89
  hydrolysis of, 38, 39
  structure, 32
Amylose
  content in waxy wheat, 3
  and crumb firming, 89
  hydrolysis of, 38, 39
  structure, 32
Analysis of wheat and flour. *See* Tests
Antioxidants, 35
AOAC International
  analytical methods of, 47
  Web site, 64
Arabinoxylans, 33–34, 37, 40
Ascorbic acid, 42, 80
Ash content
  and extraction rate, 21
  in flour, 24–25
Ash test, 21, 50–51
Aspiration, in wheat cleaning, 18
Association of Operative Millers, Web site, 24
Azodicarbonamide, 42, 80

Bagels
  ingredients, 82
  processing, 84
Baking of hard-wheat products, 86
Baking powder, 99–100, 107
Baking tests, of flour, 62, 71
Batter, 42, 101
Benzoyl peroxide, 43, 119
Biotechnology, for wheat improvement, 3, 9
Biscuits, in United States, 103, 106, 110, 112
Biscuits, outside United States. *See* Cookies
Blackpoint, 120, 121
Bleaching, of flour, 40, 43
Blends, of wheat varieties, 10–11, 16, 17
Blisters, 85, 87, 93
Bran
  and ash content, 50
  contamination of flour, 50
  flour fraction, 21
  layer of wheat kernel, 6
  in milling, 19, 21
Bread. *See also* Breadmaking
  crumb, 60, 86, 89, 93, 94
  crust, 86, 89, 93
  ingredients in, 79–82
  quick, 102, 104
Bread Baker's Guild of America, Web site, 92
Breadmaking
  enzymes in, 43, 91
  flour for, farinograph parameters, 54
  mechanism of, 91–93
Break system, 19, 23
Breeding, of wheat, 3, 12, 56
Browning, 39, 60, 87, 93

Cake-making ability, of flour, 43
Cakes
  angel food, 101, 102
  crumb, 107
  layer, 101
  pound, 102
  problems, 107–108, 111
  processing of, 103–104
Calcium iodate, 42, 80
Calcium peroxide, 42, 43, 80
Calcium propionate, 80, 90
Caramelization, 87, 88, 93
Carotenoids, 35
Checking (product fault), 109, 118, 120, 121
Chicago Board of Trade, Web site, 13
Chinese noodles, 116
Chlorination, 43, 51, 97
Cleaning, of wheat, 17–18, 23
Clear flour, 21
Clostridium perfringens, 62
Club wheat, 2, 4, 5
Color
  of baked products, 87
  of flour, 21
  of pasta, 116, 119
  problems, 93
  of semolina, 35, 40
  tests for, 48–49, 72
  of wheat, 2
Combustion methods, for protein content, 51
Common wheat, 2
Cookies, 103
  problems, 112
  processing, 105–106
  spread, 58, 109–110
Cooling, of products, 86
Crackers, 102, 104, 109, 112
Cream of tartar, 99
Crop-year changeover, 70, 74–77

Crystallinity, of starch, 32
Cutoff flour, 21
Cysteine, 29, 31, 42
Cysteine hydrochloride, 40

Damaged starch, 22, 37, 43, 63
Denaturation, of proteins, 29, 107
Dextrin, 38, 39
Diacetyl tartaric acid esters of monoglycerides, 80
Dicalcium phosphate, 82, 99
Dies, for pasta, 118
Docking, 84
Dough
- additives, 57
- changes during development and baking, 91–93
- conditioners, 80
- gas-holding ability, 42, 42, 91
- overmixing, 53, 54, 84
- and relaxation, 84, 90
- sticky, 82
- strengthening, 41, 82
- sweet, 82
- tests of, 52–58
- work in put, 51, 90, 103

Dough development time, 53, 56
Doughnuts, 102, 104, 108, 112
Dry milling
- cleaning, 17–18, 23
- mill systems, 18–21
- products, 21
- of specific wheat types, 22
- tempering, 18, 22, 23

Dry stoner, 18
Drying
- of pasta, 118
- of wet-milled products, 24

Durum wheat, 2
- growth regions, 4
- milling of, 22–23
- products made from, 5, 115–122
- specifications for, 72
- thousand-kernel weight, 48

E. coli, 62
Egg whites, 101, 107
Eggs, in noodles, 116
Einkorn wheat, 2
Emmer wheat, 2
Emulsifiers, 100
Endosperm, of wheat, 6, 34
- in milling, 19, 22

Enrichment, 41, 44, 120, 121
- tests to detect, 52

Enzymes. *See also individual enzymes*
- and amino acids, 31
- analysis of, in flour, 59–61
- and ash content, 50–51
- in dough, 43, 91
- in germinating wheat, 7, 8

Ethoxylated monoglycerides, 80
Exporters of wheat, 9–10
Extensibility, of dough, 56, 57, 58
Extensigraph test, 56–57
Extraction rate, 21

Falling number test, 59–60, 71, 72
Farinograph test, 52–55, 72
Fat, in pie crusts, 106, 110
Fat absorption, by doughnuts, 108
Fermentation, 83, 84, 91
Filled products, 84
Firming of crumb, 88–89, 94
Flat breads, 82, 84, 86
Flavor
- from browning, 84
- from malt, 43
- problems, 94
- and proofing, 85
- undesirable, 40, 88. *See also* Off-flavor

Flour
- additives, 41–43
- agglomeration of, 49
- all-purpose, 43
- bread, 43
- consistency, 67–68
- enzymes in. *See* Enzymes
- high-ash, 24–25, 49
- high-protein, 82
- lipids, 34–35, 38
- miller, priorities, 24–25
- nonstarchy polysaccharides. See Nonstarchy polysaccharides
- particle size, 22
- protein. *See* Protein, in flour
- in soft wheat products, 97
- specifications for. *See* Specifications, for flour
- sprout-damaged, 116
- starch in. *See* Starch
- storage of, 23
- strength, 24, 58
- testing of, 48–63
- types, and extraction rate, 21
- water requirements, 59
- white, 44
- whole wheat, 21, 44
- yield, 25, 48

Flour dressing system, 20–21
Foam, 101
Folic acid, 41
Formulation, of products, 101–103
Fusarium, infection of wheat, 8

Germ, of wheat, 6
- in milling, 19, 21

Germination, of wheat kernel, 6–7
Globulins,, 27, 30, 31
β-Glucans, 34, 37
Gluco-δ-lactone, 82, 99
Glucose structure, 31
Glutamine, 29, 30
Glutelins, 28
Gluten
- agglomeration of, 108
- and crumb firming, 89
- functionality, 35–36
- in hard-wheat products, 80, 82
- separation from starch, 23–24

Gluten complex, 36, 92
Gluten proteins, 22, 31
- gliadin, 28, 29, 30, 35–36
- glutenin, 28, 29, 30, 35–36, 42

Gluten washing tests, 58
Grades of wheat, 11
Grain Standards of the United States, 11
Grainnet, Web site, 24
Gravity table, 18
"Gray dough," 41

Hard wheat
- flour from, 35, 43, 58
- growth regions, 4
- growth time, 7
- milling of, 22
- products from, 5, 79–95
- specifications for, 72
- starch damage in, 63
- tempering, 18, 22
- thousand-kernel weight, 48

Hearth breads, 83
Helix, 29, 32
Helminthosporium sativum, 120
Hemicellulose, 33
High-protein wheat, defined, 17
Hydrochloric acid, 43

Identity preservation, of wheat varieties, 10, 11
Importers of wheat, 9–10
Ingredients
- in durum-based products, 115, 116–117
- in hard wheat products, 79–82
- number of, 81–82
- premixing, 83
- in soft wheat products, 97–101

Insects
- and growing wheat, 8
- and storage of flour, 23, 49

Iron, 41, 52, 120

Kansas City Board of Trade, Web site, 13
Kernel, wheat
- hardness, 22, 48
- structure, 5–6

Kjeldahl test for protein, 51

Lactic acid bacteria, 88
Lamination of dough, 84, 90
Lard, 103, 110
Leavening
  chemical, 98–99
  gasses, 36, 39, 43, 79, 85, 91, 92
β-Limit dextrin, 39
Linkages, glycosidic, 31, 32, 33, 34, 38, 39
Lipases, 40
Lipoxygenase, 40, 59, 73, 117, 119
Low-grade flour. *See* Clear flour
Lysine, 44

Magnesium, 40
Maillard browning, 60, 84, 87, 88, 93
Malt, 43, 60, 72, 80, 88
Maltose, 39, 43
Microbes, 23, 49
Middlings, 20
Millfeed system, 21
Milling. *See also* Dry Milling, Wet Milling
  history of, 15–16
Minerals
  binding of by phytate, 40
  content in flour, 21
Minneapolis Grain Exchange, Web site, 13
Mixes
  preassembled, 102
  prepared, retail, 104
Mixing, of dough and batter, 83, 103, 104, 105, 106, 117, 119
  overmixing, 53, 54, 106, 108
Mixing curve, farinograph, 53
Mixing tolerance index. *See* Tolerance index
Mixograph test, 55–56
Moisture, and wheat crop growth, 8
Moisture content
  conversion of, equation, 50
  in dough, 40
  in pasta, 118, 120
  of stored flour, 23, 49
  of stored wheat, 12
  tests for, 49–50
Molds
  and growing conditions, 8
  and storage conditions, 23, 49
  on surface of bread, 80, 89–90
Monocalcium phosphate, 99
Monocalcium sulfate, 82
Muffins, 102, 104

National Baking Center, Web site, 92, 111
National Pasta Association, Web site, 121

Near-infrared reflectance spectroscopy, 51, 63
Niacin, 41, 52
Nitrogen content, conversion to protein content, 51
Nonstarchy polysaccharides, 33–34
Noodles, 115
  ingredients, 116–117
  problems, 120–121, 122
  processing, 118–119
Nutritional aspects of wheat, 44

Odor, in flour, 23, 38
  test for, 49
Off-flavor, 38, 40, 81
Oil, for frying, 102, 117
Oriental noodles, 119
Overmixing, 53, 54, 106, 108
Overproofing, 85
Oxidants, 41–42, 80

Pancakes, 102, 104
Particle size
  of flour, 22
  of semolina, 23
Particle size distribution, of flour, 22
Pasta
  ingredients, 116
  problems, 119–120, 122
  processing, 117–118
  types, 115
Pasting, of starch, 36
Patent flours, 21, 41, 49, 51
Peak consistency, 55, 105
Pekar color (slick) test, 48, 71, 73
Pentosanases, 40
Pentosans, 33, 34
Pests, and wheat storage, 12
pH level, 43
  in batters, 108
  and chlorination, 51–52, 97
  test for, 51–52, 72
Phenolic compounds, 41
Phytase, 40
Phytate, 40
Pie crusts, 103, 106–107, 110, 112
Pigments
  in flour, 35, 40, 41, 43
  in products, 92, 117
Pita bread, 82
Pizza crust, 82, 84–85
Pleated sheet structure of proteins, 29
Pocket bread, 82, 86
Polyphenol oxidase, 41, 59, 116, 121
Potassium bicarbonate, 82, 98
Potassium bromate, 42, 80
Potassium iodate, 42, 80
Potassium sorbate, 42, 80–81
Preservatives, in products, 80, 82

Pretzels, 82
Processing
  of durum-based products, 117–119
  of hard wheat products, 83–87, 90
  problems, 94, 107–110, 119–121
  of soft wheat products, 103–107
Proline, 29, 30
Proofing, 85–86
Propylene glycol monostearate, 80, 108
Proteases, 39–40
Protein, in flour. *See also* Gluten proteins
  content
    and breadmaking characteristics, 51
    tests for, 51, 63
  Osborne classification system, 27, 28
  molecular weight, 29, 30
  quality, 52
  structure, 28–29
Protein, in wheat
  content and wheat yield, 8
  test for, 63
Pumpernickel bread, 86
Purification system, 23
Purifiers, 19

Quality, of flour
  communication about, 68
  defined, 67
  specifications for, 69–71

Rancidity, 41
Rapid Visco Analyzer, 60–61, 72
Reducing agents, 42
Reduction system, 20, 23
Regulations, federal, 69
Relative humidity, and proofing, 85
Resistance to deformation, 58
Resistance to extension, 56, 57
Respiration, of stored wheat, 12
Retailer's Bakery Association, Web site, 92, 111
Retrogradation, of starch, 36
Riboflavin, 41, 52
Roller mill, 18, 22
Roll-in fat, 84, 90
Rolls, in mill, 19, 20
Run chart, 73

Salmonella, 62
Salt, 79, 88, 98, 116
Scaling, 83
Scourer, 18
Semolina
  color, 35
  granulation test, 52, 73
  particle size, 23
Separators, 18
Sheeting, 90, 119, 121

Shelf life, of products, 88–90
  problems, 94
Shortening, 80, 82, 100, 103, 110
Shorts, 21
Sieves, milling, 19, 22
Slicing, 86
Sodium acid pyrophosphate, 99
Sodium aluminum phosphate, 82, 99
Sodium aluminum sulfate, 82
Sodium bicarbonate, 82, 88, 98, 99
Sodium metabisulfite, 40, 42
Sodium pyrophosphate, 82
Sodium stearoyl lactylate, 80
Soft wheat
  defined, 2
  flour from, 35, 43, 58
  growth regions, 4
  milling of, 22
  products from, 5, 97–112
  specifications for, 72
  starch damage in, 63
  tempering, 18, 22
  thousand-kernel weight, 48
Software, for testing, 54, 56
Solvent retention capacity profile, 59, 72
Sorbate, 40
Sorbitan monostearate, 80
Sourdough bread, 88
Specifications, for flour
  changes in, 67, 70, 77
  and crop-year changeover, 74–77
  information to include, 69–71
  for specific types of wheat, 72–73
  testing procedures used, 71–72
  use of, 73–74
Specks, bran, 23, 48, 73, 97, 103, 120, 121
Sponge-dough process, 83–84, 102
Spring wheat, defined, 2
Sprouted wheat, 7, 8, 38, 59
Stability, mixing test parameter, 54, 56
Staling, 88, 89
Staphylococcus aureus, 62
Starch
  damage, 25, 48, 63, 72, 97
  functionality, 36–37
  gelatinization, 36, 87, 92, 101, 107
  gelling, 107–108
  isolation of, 23–24
  structure, 31–33
  tests of, 62–63
  wet-milling product, 24
Starch granules, 22, 32–33, 92
Steam
  in baking oven, 86, 87
  for leavening, 82
Stirring number, 61
Storage
  of flour, 23, 49
  of wheat, 12
Straight-dough process, 83, 102
Straight-grade flour, 21
Sucrose
  and delayed gelatinization, 100
  as plasticizer, 100
Sucrose monostearate, 80
Sugar
  and flavor, 88, 100
  in products, 101, 102
  as sweetener, 80, 82
  and yeast fermentation, 80, 91
Surfactants, 102, 108. *See also* Dough, conditioners

Temperature
  for baking, 86
  for cooling, 86
  of enzyme inactivation, 39
  of flour storage, 23
  for proofing, 85
  of starch gelatinization, and α-amylase, 38, 59
  of tempering, 18
Tempering, 18, 22, 23
Test weight, 25, 47
Tests
  of flour performance, 52–64
  of wheat, 47–52
Texture, of baked products, 87, 93
Thiamine, 41, 52
Thousand-kernel weight, 25, 47–48
Tocopherols, 35
Tolerance index, 54
Tortillas, 82, 84
Triticum
  aestivum, 2
  compactum, 2
  durum, 2
Troubleshooting, 93–95, 111–112, 122

Ultraviolet irradiation, for enrichment detection, 52
United Nations Food and Agriculture Organization, Web site, 13
U.S. Durum Growers Association, Web site, 121
U.S. standards for wheat, 11, 17
USDA Foreign Agricultural Service, Web site, 13
USDA, National Agricultural Statistics Service, Web site, 75, 77

Varieties, of wheat, 2–3
  identity preservation, 10, 11
Vernalization, 2
Viscoelastic properties of dough, 57, 58, 121
Viscosity
  and α-amylase, 59, 60, 61
  and chlorine, 43
  and starch gelatinization, 36–37, 60
  tests for, 60, 61, 72
Vital wheat gluten, 24, 82
Vitamin C, 41

Wafers. *See* Cookies
Waffles, 102, 104
Water
  binding of, 36–37, 40
  in bread formula, 80
  relationships, in dough, 37, 58, 75, 97, 102, 103
Water-binding ability
  of new-crop flour, 75
  of starch, 39
Waxy wheat, 3
Web sites, 13, 24, 44, 64, 77, 92, 111, 121
Weed seeds, 8
Wet milling, 23–24
Wheat
  classes, 2
  crop tour, 75
  grading, 11, 47
  growth regions, 4–5, 16, 76
  growth stages, 6–7
  hardness, 63
  harvest data, 16, 75–76
  history, 1
  marketing, 9–10
  producer, priorities, 12
  production, 8–9
  protein. *See* Protein, in wheat
  soundness, 25, 47
  sourcing, 16–17
  standards, 11, 17
  storage, 12
  testing of, 47–48, 49–51, 63
  types, 2,3
  yield, 8
Wheat Quality Council, 75
  Web site, 77
Whole wheat flour, 21, 44
Winter wheat, defined, 2
World Grain, Web site, 24

Yeast
  during baking, 92
  during fermentation, 79, 83, 91, 109
  and flavor, 88, 102
  levels in products, 79, 82
Yeast food, 80
Yield
  of flour, 25, 48
  of wheat, 12
    and protein content, 8